JUST TRAVELING

JUST TRAVELING

God, Leaving Home, and a Spirituality for the Road

JACO J. HAMMAN

FORTRESS PRESS
Minneapolis

JUST TRAVELING
God, Leaving Home, and a Spirituality for the Road

Cover image: Photo courtesy of Michnus and Elsebie Olivier (www.pikipikioverland.com). Taken outside the Patagonian village of El Chalten, Argentina, and next to the granite peaks of Cerro Torre and Mount Fitz Roy.
Cover design: Naughton

Print ISBN: 978-1-5064-7206-5
eBook ISBN: 978-1-5064-7207-2

To all with the courage to leave home

CONTENTS

PREFACE

Just Traveling has been hundreds of years in the making, for I carry the DNA of Johann Jürgen Hamman, a metalsmith who left Staucha (near Dresden, Germany) to arrive in Cape Town on a Dutch merchant ship, the *Petronella Alida*, in 1732. As a child, my imagination around his journey ran wild. That my family loved road trips, camping, and exploring was unavoidable. We left our home often, especially for short trips, as every Sunday afternoon we would go for a ride, first to explore areas around Johannesburg and later to enjoy the beauty of South Africa's southeast coast. June and December brought longer trips to national parks and family. This love for open roads and new experiences empowered me in 1993 to leave South Africa in an effort to avoid military service during the apartheid administration by pursuing chaplaincy education. The United States, which gave me a second passport, offers countless miles, places, and parks to explore. The past years, I have crisscrossed the United States and Canada, more than 150,000 miles on a motorcycle alone. I also travel to teach, to learn, and to enjoy family vacations. Along the way, I blog and keep notes of my experiences, knowing that the privilege of traveling demands deeper reflection. Some of those experiences appear in this book to point not to me but to humanity's traveling spirit and hospitality.

The importance of traveling deepened as I aged. In the early 2000s, it quickened as I found myself in a system where many individuals were physically sick or emotionally stressed. It was as if the anxieties of the academic and ecclesial worlds that surrounded us seeped into human bodies. Seeking a way to escape the system, I bought my first motorcycle at age thirty-five. We travel, I have come to believe, in part to reclaim personal health and well-being and also to work on ourselves. Furthermore, traveling convinced me that one's personal well-being is inextricably tied to the well-being of others and the vitality of environments and the earth. We certainly travel for many reasons, including very personal ones, but we always travel in a beautiful, complex, and troubled world. Remaining aloof and apathetic to this world goes against the traveling spirit and cannot sustain travel as a life-giving and transformational practice. To travel is thus never *just* traveling; it beckons one to contemplate personal, social, financial, even political and ecological realities and values.

I wrote *Just Traveling* while the world sought isolation and protection from the COVID-19 pandemic. Reflecting on the traveling experience, knowing that leaving home is coded into our DNA, and finding solace that a pandemic will not eradicate the traveling spirit were sources of joy and hope in uncertain times. They reiterated the fact that one can travel through the gifts of imagination, anticipation, and engagement without physically leaving one's home. Writing about travel nurtures the traveling spirit but is not easy.

"There are fundamentally two kinds of writer," sage Adam Phillips writes, "the immaculate and the fallible. For the immaculate every sentence must be perfect, every word the inevitable one. For them, getting it right is the point. For the fallible, 'wrong' is only the word for people who need to be right. The fallible, that is to say, have the courage of their gaucheness, they are never quite sure what might be a good line, and they have a superstitious confidence that the bad lines somehow sponsor the good ones."[1]

If this is true for writers, and I think it is, then it is certainly true for travelers too, where one group of travelers is open to failure and the uncertainty of open roads. By *writing* about traveling, I may be revealing more than I want to or am comfortable with yet. The traveling experience, however, is rewarding enough to take the risk.

May this book be a conversation that can take a reader places.

Jaco J. Hamman
Nashville, August 2020

ACKNOWLEDGMENTS

I am grateful for all the places and people who welcome travelers with hospitality, patience, and a transformational spirit.

My partner and our daughters allow me to travel for extensive periods on my own. Their grace when I become absent-minded or border on obsession is supportive and permission-giving. Our joint travels make life worth living.

Carey Newman remembered a conversation, one that occurred in passing, and saw potential in a book exploring the spirituality of traveling. Savanah Landerholm's creativity became an eye-catching cover. Working with Carey, Savanah, and the 1517 Media's Fortress Press team reflected the excellence and professionalism they and the publishing house are known for.

Vanderbilt University and Vanderbilt Central Library support my research and writing in essential ways. I am especially indebted to Chris Benda in the Divinity Library. Chris not only helped me secure resources (alongside the library's interlibrary loan service) but also pointed me to resources and provided feedback on each chapter.

Michnus and Elsebie Olivier, in their eleventh year of motorcycle travel at the time of drafting this book, sponsored the cover photo. Their blog allows many, including me, to travel vicariously and be inspired.

Vanderbilt University doctoral candidate Rev. Christopher Hoskins served as a research assistant and conversation partner on this project.

God's self-revelation as the Earthroamer surprised me. Never will my understanding of God be the same again.

TRAVELING

A Paradoxical Practice

> We are inundated with advice of *where* to travel, but we hear little on *why* and *how* we should go, even though the art of travel seems naturally to sustain a number of questions neither so simple nor so trivial.
>
> —Alain de Botton, *The Art of Travel*

To travel and to explore are defining features of the human spirit. Traveling is a practice that invites contemplation, awakens curiosity, and reminds us that we are part of a much larger whole. In lonely moments away from home, we can discover a mysterious sense of belonging—such is the paradoxical power we'll explore in this book. Much of what follows first took form away from home. The mental notes grew into clip notes, which became words on paper as thoughts gathered from across the globe warranted deeper reflection.

"Highway 2—it's always good for lovers, always good for poets, and people passing through," the artist sings.[1] The song suddenly had new meaning as I was about to ride "Highway 2" west to east. The road was the chosen route between Blaine (Washington) and Madawaska (Maine), the westerly and easterly most northern corners of the United States, 3,250 miles apart. A couple overheard my conversation with a Blaine coffee shop hostess and asked, "Why would you ride a motorcycle from Blaine to Madawaska?"

"There is so much to see and experience in this country," I answered, my accent long since having given me away as a stranger.

Each one of us has our own reasons for traveling, but all travelers know that roads provide what we long for.

US Route 2 is 2,571 miles long and crosses eleven states east-west in two sections, bifurcated by the Great Lakes and Canada dipping into the United States. In Washington, it traverses the Cascade Mountains, a final frontier. Bear warnings remind all they have left the city. Fly fishers line the Priest River in Idaho. Their tai chi–like cast motions and flies hooked into fishing vests awaken Zen-like peace in passersby. Traffic slows down in the hope to see a fisher land a trout before speeding off. In Montana, Route 2 takes you through Glacier National Park, where peeling off on the Going-to-the-Sun Road offers exhilarating drop-offs, breathtaking views, surprising wildlife, and retreating glaciers. Entering North Dakota and heading to Grand Forks, the road becomes a four-lane highway, courtesy of the omnipresent oil and natural gas industries. One passes Rugby, named after the town north of London that gave us a sport of legends. Once in Minnesota, Route 2 belongs to lumber trucks, for this is the world of Paul Bunyan and Northwoods mythologies. Approaching Wisconsin's Superior reminds you that you are next to the Great Lakes. And then one enters Michigan's "UP," the state's Upper Peninsula, where Finnish, Swedish, and French Canadian settlers joined Algonquian-speaking Native American peoples. Stopping in Marquette strains one's ears as Yoopers speak "eh" dialect that is difficult to decipher. It is 1,120 miles to Madawaska if one cuts through Canada, which is the obvious choice.

Many share memories of family vacations, memories that are awakened whenever they travel. While riding, I am often transported back to being a child in South Africa in the early 1970s. I remember the winter holidays, the best time to visit Southern Africa's game parks. In the previous days, family arrived from "the south" and "the farm," and a convoy of cars, packed late into the night, was parked in our driveway. The atmosphere buzzed with excitement. We knew the departure time, four o'clock in the morning, determined by the desire to pitch our tents before sunset. Despite threats from parents, sleep came slowly to us kids as the excitement of what awaited us filled our bodies and imaginations. Due to the early departure, we could sleep in our clothes. Breaking customs such as wearing tomorrow's clothes today and going to bed later than usual indicated the beginning of a new time.

There was laughter, a sense of deep belonging, feeling known and loved, and gratitude. Alarms woke us all at 3:40 a.m. as we all quietly got into vehicles packed to the ceiling. Our silence reflected this holy moment of leaving. We children would soon fall asleep, awakened for a roadside breakfast around eight o'clock. We knew what to expect: sausage grilled the night before, hard-boiled eggs, rusks, marmite-and-cheese sandwiches, and sweet coffee or tea—a complete meal. The changed landscape of syringa, fever and thorn trees, and bright-yellow grass hiding kudus and cobras told us we were entering the *bosveld*. New sounds greeted us as birds chirped, red-billed hornbills joined us for breakfast, and local communities started moving about their day. Boys who have not claimed their manhood yet played with *draadkarre*, motor vehicles with recognizable shapes made from wire and soda cans. The Volkswagen Microbus and *bakkies* were popular. As a boy from the city, I lacked the skill to build a *draadkar* and silently coveted the ones I saw. Arriving in the early afternoon at the game park, we'd set up camp. The dark in the bush is darker than at home. The stars are touchable should one reach out. The sounds of lion, hyena, and jackal made sure we kids never wandered beyond the light of the fire. We would change camps twice more with the promise of new prides, herds, and terrain. It may very well have been the very best days of my life. As a five-year-old, I knew traveling would forever be part of who I am. Questions about how African game parks were formed in the colonial era or what happened to peoples who called this their ancestral land never entered my mind. It is grace that one's awareness of justice and equity grows with age. Much has changed since the early 1970s, and fewer kids may know the joys of a road trip.

The majority of humanity—who would do the research to prove it?—knows loneliness and abandonment or at least disappointment and uncertainty. We live in a world that breeds suspicion and distrust and fuels anger, especially at strangers. A Bedouin proverb states that the stranger is also the wolf, to be feared. Self-reliance is promoted as a subtext of freedom, neglecting the ways we are interdependent as persons and dependent on nature. We still thank God for Fridays, since most layoffs occur on Tuesdays, but as we celebrate Friday, we also know that Tuesday is coming. We find ourselves on an emotional roller coaster with few uplifting sections. We know feelings of helplessness, moments of overwhelming anxiety and fear, and morbid thoughts about our individual and communal futures. We wonder whether the earth will sustain us much longer. The rich, we read, are getting richer. Many of us have barely enough cash on hand to sustain ourselves in a minor crisis. For good reason, we have lost trust that our societies and this world will deliver in expected ways. We fight hard to buffer these feelings, leaning toward despair with experiences that instill joy, contentment, and hope. Road trips and traveling are not the foremost of thoughts; concerns stubbornly lurk and refuse to retreat. Most of us remain rats on wheels, racing through life but not moving forward.

Cultural critic Camille Paglia aptly describes how we—stuck between forces of order, symmetry, and unity and forces of chaos, disorder, and disunity—skim across life. "[The] classic American experience," she writes, is "driving flat-out on a highway, radio blaring. Driving is the American sublime. . . . Our long, straight superhighways crisscross vast space. . . . To traverse or skim the American landscape . . . is to feel the speed and aerated space of Don Juan."[2] Thirty years have passed since Paglia wrote this. In the interim, time has become an even more precious commodity, so rather than driving across country, we fly. There is a hidden arrogance in thinking one can "save" time. Anxious about our lives, we skim the world in planes, crisscrossing vast spaces to experience what was promised in brochures, hoping that it will revitalize us for another session on the ever-present rat wheel. When we have packed our bags to skim back home, we wonder whether the tip we left in the hotel room was large enough for the condition we left the room in. The vague awareness that an anonymous person will clean up after us remains.

To travel is to move at the speed of being present to one's experiences, bridging distance and difference through acts of care.[3] The practice has mostly evolved past its Latin roots of *tripalium* (a torture instrument) and its French origins in *travail*, which speaks of bodily and mental labor, exertion, pain, and even hardship and suffering. We can plan it all online. Still, contemporary travel sometimes feels like work—all the planning, all the adjusting, and then all the work to catch up on when we return. To travel is to be mindful of the self, the other, and also things, recognizing those ways we seek to control and being surprised by the ways we are transformed into likenesses. The "other" is a part of you that is yet unknown to yourself but also a person, a group, a community, a city, all of nature, even

God. The "other" comes in the form of dreams, meeting a person by coincidence, walking along ancient walls, feeling your heart beating as you climb a mountain, or experiencing breathtaking vistas. There is grace that traveling anticipates meeting a stranger, whether that stranger is us or someone else. Whether it is because of our emotional, relational, spiritual, or financial anxieties; risk aversion; the general distrust we have toward the world; or skimming across the earth to tick off one more place on our wall map—traveling as a practice that promises transformative experiences is elusive, maybe even under threat. Our definition of traveling overlaps with the definition of tourism as "a social, cultural and economic phenomenon which entails the movement of people to countries or places outside their usual environment for personal or business/professional purposes," though placing tourism and traveling in a binary relationship is not helpful.[4]

Embracing Paradox

The practice of traveling is an unsolvable paradox: we seek to get away to be present, we want to slow down time but speed up to get to our destination, and we care, but in traveling, we leave carbon footprints. Traveling is cheeky, even disrespectful. It ignores our current life, reminding us that it is always a good time to travel. It overlooks social norms of gender, age, different abilities, and an arbitrary category such as distance. Anyone can travel, even if only in one's mind, though travel can be a significant challenge for persons with diverse abilities and may look different for persons with no financial resources. Tying traveling to wealth, privilege, and a neoliberal capitalist market economy depends on one's definition of traveling as a meaning-making practice. Travel as being present to one's experience does not depend on one's financial resources, for placing oneself in a position to experience something new is even possible for those with limited resources or mobility. Neither place nor prosperity defines travel, for traveling is measured differently. There is little excuse not to travel.

Besides being disrespectful, traveling is also selfish and demanding. You are the one leaving as loved ones and friends are staying behind, or your mind is expanded in ways others do not notice or understand. Traveling has interests and endurance beyond what is sold as exotic or with promises of short-term excitement. Skimming, though usually about speed, takes many forms. Sometimes we skim at a snail's pace as we stand in line for the ultimate buffet of a cruise ship inching toward the next harbor. The shine of novelty rubs off in time. Rather, traveling expects that one will be touched by persons and communities and nature, not always in physical ways with handshakes and hugs, but always in emotional, relational, and spiritual ways that honor the deep connection between all people and between us and nature. The traveler is hungry for experience.

Exploring travel is fraught with the possibility of disaster, especially if we compare it to being a tourist. We are warned by anthropologist Marielle Risse that "people who dispense definitions of 'tourists' reflect the worst of country club denizens deciding on members. There is a pervasive feeling that the definers have passed the Rubicon themselves and refuse to remain in the same company as those who may have merely crossed the Amazon, Mississippi, or Nile. . . . Yet despite the attempts to divide travelers from tourists, a division between the two groups is difficult, if not impossible to maintain."[5]

Just Traveling is not a snobbish attempt to suggest that what we need to know about travel is clear. Rather, the author's failures and missteps as he attempts traveling will become clear. The book is not about crossing rivers, visiting exotic destinations, having money to travel, acquiring a second or third language, exhibiting physical toughness, or believing that long distances are better than short distances. There is no "real (or true) traveler" to aspire to or "essential" traveling experience to chase. Traveling's irony is that the short distance from one's home to the other side of town may be a distance too far for many to discover, and a place yet to receive recognition is often the best place to visit. Our definition of traveling is meant to frame a conversation as we recognize a spectrum of experiences. It's not that traveling is "good" and tourism is "bad," for we need different experiences in this life and such splitting never serves a person well. Traveling and tourism and their related descriptors can best be placed on a spectrum of experiences.

Imagine a couple, Taylor and Jordan, close to what the Japanese call *karoshi*, death by overwork. They plan a Caribbean trip to rekindle their relationship, be rejuvenated, and get away from the day-to-day grind. Their island hosts see them as guests. We'll compare them to a family, the Bosch family, a family of five, including three teenagers. The Bosch family has a different agenda for their trip.[6] They save as much money as possible, even forgoing birthday gifts, because traveling is their passion. Taylor and Jordan seek to *overcome* vast distances at great speed—skimming, as Paglia indicates. From their West Coast home, they want to get to the Caribbean without delays (*chronos* moments determined by a clock). The Bosch family, in contrast, seeks to *discover* distance and look out for *kairos* moments (moments that are pregnant with possibility and that demand decision and action). Their minivan has many long-distance miles, as they prefer driving to flying. They do so not primarily as an environmental statement but to see new places and meet different people. Should they fly, the Bosch family often adds a day or two to layovers to scout a new city as part of their journey. For Taylor and Jordan, *arriving* in perfect weather is important. They prepay for *hospitality and self-indulging experiences* such as parasailing and snorkeling. The Bosch family rarely has detailed plans and sees what the day might bring. They receive *hospitality as a gift and are willing to give up privileges* and even perceived rights. Taylor and Jordan do not seem to mind their commodification and that the travel brochure objectifies near-naked women promising "fun in the sun." Poolside, they experience much through their cell phone screens as they upload photos and reflections to their social media platforms. They *look at* things between moments of *problem-free excitement*. The Bosch family seeks to build memories *with* others, has few photos to show, and never seems to be in a rush.

Taylor and Jordan sought an *exciting destination* that came with rave reviews from their friends. The *road less traveled* was not something they entertained. The Bosch family has had many setbacks while driving the backroads of America, including a previous minivan hit by a teenager new to driving. They are still friends with the family of that teenager because *care and relationships* are important to them. As Taylor and Jordan lounge by the pool, they prefer *to be left alone*. They passed on an opportunity to visit the island's heritage sites because they are *disinterested* in their surroundings. The Bosch family's fridge is a gallery of the sites and towns they have *discovered* and stimulates conversation and rich memories.

As the end of Taylor and Jordan's trip approaches, all they have exchanged is *money*, though they feel it was worth it. The Bosch family does not have money to offer and seeks to give and receive the gifts of *experiences and lives*. The days in the sun, good food, and sweet cocktails show on Taylor's and Jordan's *bodies*, transformed by a few days away. The Bosch family often returns more svelte from all their walking and hiking but with *hearts and souls* that are full.

We are tempted to think that Taylor and Jordan are tourists and the Bosch family are travelers, but such depictions should be resisted, as Risse warned us. Splitting tourism and traveling into a binary world is too simplistic. We do see, however, a spectrum of experiences and choices, which each person can make as they anticipate their travels. To travel and enjoy a select part of the world, we often must skim across many a continent and ocean first. Furthermore, there are many reasons why we travel. A brief glance at travel shows that the variations found are ever evolving: business travel, urban travel, cultural tourism, community tourism, ecotourism, slum tourism, wine tourism, coffee tourism, medical or health care tourism (seeking medical services elsewhere), justice or activist tourism, farm tourism, volunteer travel, remembrance travel (taking a trip to a place with special historical meaning both to remember and to forget), vice travel, suicide travel, reproductive travel, rural travel, botanical travel, war zone tourism, and more. Each form of travel has unique purposes, goals, and stakeholders. Some prefer taking a trip, while others travel. Some say they travel, but their actions and money serve mostly the corporations that drive the tourist industry. Few travelers will claim that they have never been tourists or that they always escape commodification.

Travel is much like the ancient parable from the Indian subcontinent of a group of blind men who came upon a new-to-them animal, an elephant. Not knowing what an elephant looked like, the men set out to inspect the elephant by touch. The person touching the trunk said an elephant is like a large snake; the person touching the ear thought it was a fan of some sort. When another blind man touched a leg, the others heard it described as a pillar, possibly a tree trunk. The sides of the elephant disclosed a wall, and the tusk was reckoned to be a spear. Travel is best embraced as a multifaceted practice.

We are going to explore a theology and spirituality for traveling by looking at *core human experiences* inherent to the practice. Holding these experiences as a mirror, we hope to discover something about ourselves, about God, about our neighbors, and about the practice of traveling itself. We will engage in a wide-ranging, theological conversation. "Theologia," which for the Greeks meant utterances or oracles of God, or words or reasoning about God, asks, *What does God say about traveling?* Or more specifically, *What is the witness of the people of God regarding travel?* The answers to these questions are varied, and much like the blind men describing the one part of the elephant's anatomy they touched, they are determined by one's vantage point. By asking first, *What are core human dynamics of the travel experience?*, we find that human experience can offer new understandings in theology. *Just Traveling* identifies God as a traveler and seeks to learn from our spiritual ancestors who journeyed with God.

Meeting God the Earthroamer

That God is a traveler is something I learned from the British singer-songwriter Chris de Burgh. I was eight. De Burgh shared something I never heard in Sunday school. His album *Spanish Train and Other Stories*, released in 1975, was banned in South Africa due to its "blasphemous" nature. Three songs in particular riled the apartheid regime: "Spanish Train," in which the Lord lost a card game against a cheating devil and ten thousand souls were lost; "Patricia the Stripper," a song that needs no explanation; and "A Spaceman Came Travelling," a song imagining the star of Bethlehem as a spaceship that brought a spaceman to earth two thousand years ago. We are invited to a nativity scene and are told by the Spaceman (Jesus) not to be afraid and to have peace, goodwill, and "love for the child."[7] The apartheid leaders, many of whom had close ties to the Dutch Reformed Church, could not appreciate pop culture's imagination that saw the Son of God being a cosmic traveler. Yet God as being on the move and being a visitor permeates Scripture. Our anticipation and enjoyment of traveling reflect the *imago Dei*, the image of God into which we have been created. It's not that *all God's children need traveling shoes*, as memoirist Maya Angelou writes about her autobiographical journey to Ghana in search of her ancestors; it is that all God's children *have traveling feet*.[8]

Created to move, humans find traveling sometimes feels like a human need. Psychoanalyst Adam Phillips reminds us that "needs, like beliefs, are not essences that are fixed, but develop inside a person; they are not a pre-arranged repertoire . . . , but are essentially circumstantial, inclinations of the moment and the context; and therefore needs have no known essence. They are not urgencies waiting to happen, personal imperatives ready to go off, but artefacts made in relationship."[9] Need, like traveling, is a complex dynamic as it is constructed in and by a specific culture. *The practice of traveling is not a need; it is an irresistible desire, longing, or motivation.* Every person and culture will interpret this desire differently. The Judeo-Christian tradition informs our reflection on travel.

The very first images we discover of God in Scripture are of a God who briefly dwells in darkness, a God who is on the move, a God who creates, and a God who visits. God, we read, hovered over dark waters (Genesis 1), created earth and sky, later walked in the cool of the evening breeze, and visited with Adam and Eve (Genesis 3). Though a God who dwells in darkness holds theological promise, the latter images are of special importance when exploring travel. The God who moves with people and the God who visits is God Emmanuel, God with us. The murderer Cain experiences God with us in ways Adam and Eve did not, as Cain hears he'll be a protected wanderer (Genesis 4). Later we read that God tells the foreigner Abram that God will always be with him and his descendants. This encounter renames Abram as Abraham (Genesis 17).

God on the move appears as three persons become one to tell Abraham and Sarah that she will have a son (Genesis 18). She laughs in disbelief. God visits with Leah in her infertility (Genesis 29) and surprises Moses at a burning bush (Exodus 3). Later, God travels ahead of Israel by day as a cloud and by night as a pillar of fire (Exodus 13). God on the move offers direction, though it is confusing if we think God led Israel for forty years in circles to cover a distance of maybe two-hundred-plus miles. Humor and irony, no doubt, are part of the traveling spirit. God visiting is mostly seen as a positive experience, unless God visits the sins of the parents, which still awakens fear of judgment (Exodus 20). Moses informs the Israelites, or maybe he tells himself as he was facing his death, that God will never leave or forsake them (Deuteronomy 31). Joshua is encouraged by God's presence to cross the Jordan River after Moses died (Joshua 1). Often God reminds the Israelites, "I will walk among you and be your God, and you will be my people" (Leviticus 26:12). The prophet Isaiah hears God will strengthen his right hand, a symbol of God's presence (Isaiah 41). God, of course, also visits with an almost-pregnant Mary by proxy (through an angel; Luke 1). Matthew opens his narrative by telling how God's angel visited with Joseph to inform him that Jesus shall be called Emmanuel (Matthew 1). Having completed his ministry, Jesus tells his disciples that he will be with them till the end of time (Matthew 28). Even a cursory reading of Scripture reveals God as God on the move, wherever we go. Travelers find rest in this God for, simply put, *God is the Earthroamer*!

God with us identifies a responsiveness in God, an empathetic nature that informs the very being of God and all of God's engagement with us.[10] God the Earthroamer points beyond Jesus as the embodiment of God with us. It implies that God joins other earthroamers—all with traveling feet—with compassion, companionship, and care. God is not impassible. God the Earthroamer, who is loving, caring, compassionate, and with us, reveals *pathos* as an essential trait of God, one more compelling than any human condition, such as sin or guilt.[11] God the Earthroamer is less interested in matters of sin and salvation and more compelled by our trekking and travails through life. All earthroamers—even those who seek to travel to other planets—find rest in this knowledge. *Just Traveling* explores walking through life with the Earthroamer whose pathos drives all.

Is the Bible a Travelogue?

As we explore travel, we follow the Earthroamer, who is behind us, around us, in front of us, and, through God's Spirit, in us. Theological reflection on traveling inevitably takes us to the very nature of God. We've already argued that how we move through life can be placed on a spectrum with diverse, even opposing values. Discussing traveling exposes where language breaks down, and how one defines traveling is important. It also points to ways that the Judeo-Christian tradition challenges tourism and traveling. Traveling poorly (if at all) describes Adam and Eve's *expulsion from* Eden

(Genesis 3), turning them into exiles; Abraham's *searching for the next oasis* or *walking before God* (Genesis 13 and 17); or Hagar's *displacement* as she flees abuse and seeks to protect the life of her unborn child (Genesis 21). To argue that *exodus* (or *displacement* or *diaspora*) is traveling (Exodus 12) or that Israel's *wandering in the desert* is travel (Exodus 16) sounds tone-deaf. So too modern-day traveling cannot be confused with *journeying* or *crossing borders to trade* as the Queen of Sheba did (1 Kings 10), being *forced into exile* as Israel was (2 Kings 17), becoming a *wandering* (Aramaean) *immigrant* like Abraham, or being a stateless *asylum seeker* like Ruth (Deuteronomy 26; Ruth 2). Being a traveler is not the same as *having been sent* and having a traveling purpose like Joseph, Jesus's disciples, or the first believers (Genesis 45; Luke 10; Acts 1) or *fleeing persecution* (Acts 8). Of course, God was alongside all these groups of people on the move, but their topographies vary.

All these narratives share God the Earthroamer, who created a people constantly on the move. The Earthroamer is behind some of humanity's movement and with others when war, famine, and social disaster ignite mass migration of peoples. Whether one is traveling or on the move, the Divine is part of our experience—but how? Traveling is a vocation (Latin *vocatio*, meaning "summons, a call") much like finding employment and a professional identity. There are higher purposes in traveling. "Vocation is different from profession," writes Brazilian theologian Rubem Alves. "In a vocation, the person finds happiness in the act itself. In a profession, pleasure is found not in the act but in the gain derived."[12] The German word for "vocation," *bestimmung*, is informative too, since it indicates arriving at a certain place, a final end; it refers to a determined action and a destination. We travel to bring peace and healing and serve others, much like the seventy-two whom Jesus sent out two by two (Luke 10). Not all traveling seeks higher purposes though. Sometimes we travel just to get away, serving purely personal needs. Other times we travel to witness what is happening to a people, and other times we witness our relationship with the Earthroamer to others. We go on a pilgrimage or walk a labyrinth to discover mystery anew, even though we say God is everywhere. Meaning, purpose, call, witnessing, and pilgrimage inform being on the move in ways that further undermine making a clear distinction between being a tourist or a traveler. Suffice it to say that being on the move is a higher calling.

We learn from persons who are migrating as their own worlds have imploded under economic despair, violence, war, and the effects of climate change. Those involuntarily on the move too are present to their own experience and as such fit our understanding of traveling. Still, the argument that the Bible is a travel narrative is a half-truth at best, unless traveling is understood minimally to refer to persons moving from point A to point B or persons finding themselves on a road. One can also argue that the Bible always points to place and specific encounters that demand responsiveness, more so than depicting a people on the move. Travelers resonate with moments of being a stranger or even feeling displaced but also moments that demand a responsive presence. Although overarching (macro)narratives, such as being banned from Eden, being on Exodus's Red Sea road, or entering exile with Israel do not offer direct parallels in our exploration of travel, Scripture informs our reflection.

Traveling and Rites of Passage

People on the move have been the focus of many an academic discipline. *The Rites of Passage*, the classic text by German anthropologist and folklorist Arnold van Gennep that was first published in French in 1909, informs our theological exploration of traveling. His insights about territorial passage, "which [accompanies] a passage from one situation to another or from one cosmic or social world to another"[13]—a passage that also describes the "individual [passing] from one defined position to another"[14]—point to how and why we travel. We learn our lives consist of a series of structured transitions or rites, the details of which are determined by the culture in which we find ourselves: "[Rites] ensure a change in condition or a passage from one magico-religious or secular group to another."[15] They support social stability and carry personal, communal, and religious or spiritual significance. Rites typically indicate beginnings or leaving for something new; finding oneself in a middle, sacred space; and ending or returning: "For groups, as well as individuals, life itself means to separate and to be reunited, to change form and condition, to die and to be reborn. It is to act and to cease, to wait and rest, and then to begin acting again, but in a different way."[16] To be a rite of passage, all three elements—separation, liminality, return—need to be present. When we bring rites of passage to traveling, it becomes the practice of going somewhere (in person or in one's mind), opening oneself to liminal experiences, and returning as a changed person. Traveling is thus a performance and a practice of transformation, something we do repeatedly.

In *rites of separation*, one is removed from the ordinary life, breaking with a previous phase or social status. Separation speaks of departure and disengagement, of being away from the familiar. The moment of separation can be sacralized, submitting to religious and spiritual authority. It is often a solemn moment, stiff and formal. As one separates, linear time (chronos) stops as time becomes more cyclical with an ebb-and-flow feel; social hierarchies disappear, and persons often negate their everyday life. Today, our separation is symbolized by setting our out-of-office assistant on an email platform informing others we are away from the office. Even setting up the notification can bring joy!

Rites of liminality or *transition* place the person in ambiguous spaces betwixt and between—between heaven and earth but also between the familiar and the foreign. According to van Gennep, "As one places oneself successively in this or that spot within the society as a whole, there is a displacement of magic circles."[17] These are thin places of threshold (Latin *limen*), moments where the spiritual and religious reign and where social norms are suspended. This rite describes a transitional state. The person experiences expressions of otherness and moments of exaltation. The symbolic, spiritual, or religious is engaged as one finds oneself as part of a cosmic order. Suddenly, the knowledge from back home has little value. A sense of timelessness sets in and so too a deeper sense of community, which anthropologist Victor Turner refers to as *communitas*.[18] In this liminal space, a new identity is being formed, often one with a deeper spiritual engagement.

Rites of incorporation describe the return to the familiar social world. The transformation experienced through separation and liminality ensures that one is never the same person upon one's return. This is especially true of coming-of-age transitions where a boy or girl will leave their society a teenager and return as an adult. Often one returns with new wisdom, vitality, intentions, and also feelings of animosity against the familiar, seemingly unchanged social world. One returns to join a society that remains hierarchical and where linear time rules, but often with a new role to play. Sometimes one returns with an identity that is *unfinished* and unsettled, wondering whether a new process of separation/liminality/return will bring more clarity.

Rites of Passage in Scripture

Van Gennep describes an archetypal experience, and it is no surprise that many narratives in Scripture reflect the pattern of a rite of passage. Scripture tells of a people on the move, separation, liminal moments, and a changed return. Three brief examples will suffice: Hagar's displacement, Moses's experience with a burning bush, and Jesus's sojourn in the desert.

Hagar, the Egyptian slave girl, became entangled in Sarai and Abram's yearning for a child. We read that Sarai gave Hagar to Abram to sleep with. He did and was surprised by his own fertility. Conflict ensued between Hagar and Sarai when Hagar became pregnant. Sarai abused Hagar, who ran away. Her separation thus came as she protected herself and her unborn child. Hagar fled to a spring in the desert, where her liminal moment occurred. An angel of God appeared to Hagar and told her to return to Abram, to put up with the abuse she received from Sarai, and promised she would have a son, Ishmael, who would be blessed yet at odds with his family. Hagar returned, and Ishmael was born. However, after Sarah (Sarai was renamed by God) birthed Isaac, Sarah grew jealous again and sent Hagar, now named "Hagar the Egyptian," away (Genesis 21). Though personally distressed, Abraham (Abram was renamed too) sent Hagar and Ishmael away with some water and food. While Hagar and Ishmael wandered in the Desert of Beersheba, their water ran out. As Hagar sobbed in fear that Ishmael might die, an angel of God appeared to show Hagar a well, from which they drank. Again, she received the assurance that Ishmael would create a great nation.

Racial, economic, social, and political readings permeate this story. Biblical scholar Phyllis Trible, reflecting on Hagar, writes that "she is the faithful maid exploited, the black woman used by the male and abused by the female of the ruling class, the surrogate mother, the resident alien without legal recourse, the other woman, the runaway youth."[19] Hagar is also a woman fleeing in fear, the first person in Scripture to encounter a messenger of God in a liminal moment, and the only woman to receive a divine promise of descendants. She returns a different person, and her return fills us readers with ambivalence and questions as she returns to her abuser.

Moses experienced his liminal moment at a burning bush after fleeing Egypt (Exodus 3). His killing of an Egyptian who abused an Israelite slave was witnessed; fearing for his life, Moses ended up in Midian, where he married Zipporah, the daughter of Jethro, a priest. Moses was wandering in the desert tending the flocks of his father-in-law when he came to Horeb, known as God's mountain. Suddenly, Moses saw flames blazing out of the middle of a bush, the manifestation of an angel of God (who later became God). He was amazed that the bush was not consumed by the fire. The angel then called Moses from the bush, reminding him he was on holy ground. Standing barefoot, Moses is commanded by God to lead God's people out of Egypt. Doubtful, Moses is only convinced by God when his staff becomes a snake and then a staff again. Protestations by Moses of not being a good communicator failed too. Leaving the bush, Moses returns to Jethro to tell his father-in-law he will no longer tend flocks but instead return to Egypt. Moses's separation, liminal moment, and return take him from being a murderer-turned-shepherd to being Israel's Exodus leader.

Some of Jesus's experiences, too, reflect van Gennep's structure. Jesus, still wet from his baptism, is led into the wilderness by the Spirit, where Jesus fasts for forty days (Matthew 4). Here the liminal appearance is not a messenger sent by God, but the Devil, one of God's fallen angels. Hungry after his fast, Jesus is tempted by the Devil to turn stones into bread. He is encouraged to jump off the temple roof and to receive all the empires on earth if only he would worship Satan. Jesus is not tempted, and as the Devil disappears, angels come in to tend to Jesus's physical needs. Leaving the wilderness, Jesus returns to Galilee with a new identity as the Son of Man, ready for his public ministry of teaching and healing but not recognized by his own people.

Traveling's Rites and Practices

Rites of passage describe people on the move, whether going through the life cycle, experiencing a life or social status transition, going on vacation, or traveling. Our "interest [too] lies not in the particular rites but in their essential significance and their relative positions within ceremonial wholes—that is, their order."[20] The "essential significance" of traveling as it informs our exploration of traveling is best described if we expand van Gennep's trilogy to six distinct practices: anticipating, leaving, surrendering, meeting, caring, and returning.[21]

Anticipating acknowledges that the practice of traveling, for many, begins weeks or months, sometimes even years, before the actual travel. The act of looking forward to something and the nature of the Earthroamer are intertwined, whether it is looking forward to the promised land, the end of exile, or the final manifestation of God's reign. This anticipation is empty, however, as we do not know exactly what lies ahead or when arrival occurs. Anticipation is a powerful practice that functions in a hope-like manner as it pulls us into a looked-for future. We anticipate, of course, not only good things. Sometimes we anticipate with fear and dread.

Leaving is a practice we have honed since infancy when, crawling on our tummies, we sought ways to get away from our mom, dad, and all who held us. Leaving builds on separation as a societal need but is mindful of the emotional reaction within an individual or family at the moment of leaving. Since leaving is a common cultural expectation, not only for males, and families find their members dispersed to every corner of the world, leaving is accepted and not always appreciated for its significance.

Surrendering too is a practice informing traveling and highlights the self's longing for moments when one's whole being is fully present to an experience. Before it becomes a spiritual virtue, surrendering is an emotional one. The emptiness of anticipation compels a traveler to surrender to whatever experiences may be encountered upon leaving home. We surrender not only to the unknown; we also surrender our own desire for control. Surrender requires trust, in self, in others, in nature, in the fact that the Earthroamer is with us. Liminal moments, per definition, invite us to surrender as they confront us with something beyond our comprehension that makes us revisit personal meaning and purpose. Surrendering then is a practice and also an attitude describing the condition of one's soul. It is an attitude best embraced at the outset of one's travels. This moment requires vulnerability and comes at a cost, the loss of the illusions of control and independence, as it prepares and liberates.

Another foundational practice of traveling is *meeting*, the invitation to discover self and other anew or possibly for the first time. Meeting is an invitation to grow beyond one's preconceptions and prejudices and beckons us beyond exoticism or Orientalism, those patronizing Western attitudes and values we bring to all parts of the world. Moments of authenticity and deep connection inevitably lead to recognizing and naming one's ignorance or "confessing" those partialities we hold on to that silently fuel injustices. This acknowledgment is not necessarily a religious confession, as the inclination to confess is inherently human and not religious. When we are blinded by our preconceived notions, moments of meeting are foreclosed of their liminal potential.

Caring is an important practice for travelers, as it explores the ethical values best embraced when traveling. Caring celebrates traveling in a world where poverty abounds, where women and children remain abused, where corruption reigns, where the earth is exploited, and where forms of injustice and inequity are prevalent. With the Earthroamer, we travel differently, mindful of carbon footprints and calls for solidarity. Caring and having compassion while traveling are possible as we incarnate the Earthroamer, who is compassionate.

A final practice, which echoes van Gennep, is *returning*. It explores reentry or incorporation—that is, how to step into a familiar home, familiar relationships, even a familiar role, but to do so as a changed person. We explore what might have changed, with one's identity having anticipated and surrendered, having met people and encountered nature, and having witnessed and responded to some of the injustices of this world. There is grace in the fact that our identities are fluid and change over time. Returning reflects on how to remember and what to tell. It beckons us to benefit our wider circles of influence, family and friends, our work spaces, communities, and societies. The rites and practices informing traveling show that our travel has significance. The practices of *Just Traveling* arise from "below" as it is rooted in being at and engaging the margins.[22]

Our opening quotation by the British philosopher Alain de Botton describes travel as an art. It is a practice that includes an apprenticeship. The more we travel, the better travelers we become. Knowing why and how best to travel can help. There are many questions around traveling, de Botton writes: "one question revolves around the anticipation of travel and its reality."[23] We'll explore this question next.

Chapter 1

ANTICIPATING

Planning, Expecting, and Leaving Home

> The consideration of the future that constitutes *forethought* is a mixture of hope, faith and fear, the sum of which is the stimulant to action and progress, hope and faith being the civilized or divine motives, and fear being the animal motive.
>
> —Horace Fletcher, *Happiness as Found in Forethought Minus Fearthought*

Traveling is an itch requiring a scratch; it is getting a bug. Sometimes it is restlessness and a longing; other times it is being discontented and frustrated; and then there are times when the lure of the foreign and distant is overwhelming. Regardless, traveling is a cure. For me, the itch is often awakened by reading motorcycle travel narratives, a favorite pastime. It first appears twelve to sixteen months in advance, though I have been surprised by it weeks and days before going on a trip. Seeing traveling as taking a "trip," however, does not do it justice. My body asks for traveling, and my soul agrees with such certainty that one cannot but heed the call to go, to leave home. I keep an American Automobile Association membership for this reason only, for the balm that soothes the itch is maps. I visit the local branch to pick up maps and travel books. The tactility, smell, and sound of opening a brand-new map are sensuous. Since I will unfold the maps many times in the weeks and months ahead, I "laminate" the maps with strips of clear packaging tape. It is more flexible than other forms of lamination. Using physical maps in the digital age may perplex some, but leaning over a map at a truck stop, on a picnic table, or in a tent or budget hotel room at night is much more rewarding than scrolling on a small screen. Maps are also great conversation starters. They instill the impression that one has come some distance, but they situate you, as if placed in a specific spot. Traveling without maps is certainly possible, but not recommended. "A map," psychologist Nicholas Samstag reminds us, "is a metaphor for adventure, intended for a potential traveler equipped, perhaps, with nothing more than an urge for going. . . . A map is never a story about boundless possibilities, but always a story about possibilities within bounds."[1]

Despite having a handful of maps and books, I do venture online, reading travelogues, blogs, and forums dedicated to traveling, and I also visit online travel brokers, all in the hope of finding the best roads, discovering sites and nature's hidden gems, and locating local eateries. I ensure that my GPS has the latest software, and I keep careful notes. A future begins to take shape. On rare occasions, I reserve lodging, especially if holiday revelers in a sought-after spot may minimize the chances of my finding a bed when arriving late. The rituals of searching maps and online have sustained me since the age of ten. As a *penkop* in the *Voortrekkers*, I went on my first multiday hikes in areas of South Africa new to me. As a teenager, I left this Christian nationalist organization, but the practice of hiking and traveling over holidays stayed with me in high school and college as I backpacked in southern Africa and Europe during breaks. It has carried me through the coldest of winters, the most challenging of times, and periods when the longing for solitude overcame any needs for togetherness. Multicolored roads on maps and travel narratives richly decorated with photos pull me into a future I trust. Traveling indicates a life that resists stagnation, anxiety, and even death. Some people travel alone, some with partners and friends, and some travel in their minds and imaginations. Books and articles, travel channels, and nature documentaries take us on journeys too. We travel the distance of a labyrinth, across cities and rural areas, into nature. One of the biggest distances to cover in contemporary society is the distance between the inner city, suburbia, and rural communities. Traveling these locations exposes different values, different opportunities, different relationships with the earth and even with guns, for rural, farm-based communities often view hunting differently than folks who have never left the city. The ways of meeting one's traveling needs are diverse, yet the reward is similar. The promise of traveling is that it always takes you away, whether in person or vicariously through someone else's lens or story.

Maps are great, but they are also overrated as an antidote for the traveling itch. Maps, online searches, and documentaries serve *anticipation*, the traveler's womb. When it comes to travel, one anticipates first, then leaves. The Dutch call it *voorpret* (pronounced FOHR-pret; before or prior to + pleasure)—the joy one finds in planning and preparing. *Voorpret* resists easy English translation. When *voorpret* joins another evocative word, *wanderlust*, the German word describing the irresistible desire to travel and explore foreign places, the practice of traveling is off to a dynamic start. Itching to travel, peeking over maps, and planning ahead are symptoms and a part of *voorpret* and *wanderlust*.

The practice of anticipating is powerful and in our bones. It is also filled with possibility. Maps and notes are inherently flawed because they are fixed. The opinions and reviews on websites are similarly burdened—they present *someone else's* experience of a place and time that is locked in the past. The experience will be different for each of us because each of us is unlike anyone else; even if we want the experience to be the same, that is not possible. Traveling resists rigidity, and travelers know to hold on loosely to plans made. No seasoned traveler will ignore the advice of a local guide pointing them in a new direction. We know that sometimes we need to linger longer in a locale than planned, maybe because of relationships that have been formed, beauty, or circumstance. A traveler does not foreclose all possibility of surprise. Samstag rightly warns that "[a map's] primary conceit, of course, lies in the fact that it suggests linear, straightforward relationships of time and distance, articulated by particular biases of topography and scale, whereas the impact of any trip will likely result more from the unforeseen, the unexpected, the surprising."[2]

Anticipation, as powerful as it is in pulling us forward, is not a strategy. Whereas anticipation tends to be passive, planning and preparation are activities that increase the likelihood of satisfaction. International travelers update their passports to be valid for six months after their return; other travelers change the oil and tires and service their car, motorcycle, or motor home; they read up on border crossings; they get vaccinations for yellow fever and other diseases; they take out travel insurance; they place embassy phone numbers in their contacts; they reserve lodging, and more. Planning and preparation demand intentionality and commitment and support anticipation, much the way being aware, bracketing, and being empty do. We'll return to these practices in this chapter.

Surely the anxiety caused by rigid plans is not needed when traveling. Holding on to plans loosely is a key trait many travelers have. Some want to know the itinerary in advance, expect planes and boats and trains to leave on time, and want perfect weather and food that does not make them sick. When traveling, however, stuff happens. We know surprises are not inherently negative despite the ways our lives, world events, and even nature seek to undermine this truth. To anticipate is a human need and is foundational to our travel experience. It is a form of preparing ourselves emotionally, spiritually, and physically. We work to get our bodies in shape and plan what we will wear and take with us when we leave our homes.

Creating a packing list adds to the anticipation. What and how we pack are important, especially if planning to backpack, traveling on a small plane or motorcycle, or anticipating extreme weather. Seasoned travelers place all the items on the floor. Over days, they prune what they've set out and packed. They know they will use fewer items than anticipated. It is likely that there are people living where they are going. Finding items along the way is possible. Travelers take the best they have and resist taking new items, especially new boots or a tent, unless the boots have been broken in and the tent was pitched in the dark. They purchase additional batteries and memory cards should they plan on taking a camera, for such items are often expensive on the road or not available at all.

With the packing list pruned and the bag taking form, the moment of leaving comes closer, and the anticipation grows. Travelers find themselves and their family or housemates, including pets, acknowledging the pending departure through silence, additional touch, naughty behavior, and the odd outburst in anger. All are being prepared and sense a change is coming. Still, nothing really prepares us for leaving or being left behind, and we all experience these moments differently. I depart with much sadness. Leaving revisits trauma experienced when I was six years old and went to school for the first time. Kindergarten could have prepared me to leave mom's side, but I never attended. As I say good-bye, I begin to sob, my body shakes, and my breathing is shallow. My depths are clearly resisting any departure. Over the years, I've befriended this sobbing, which typically reappears when I arrive back home. Since some of my travel is solo long-distance motorcycling, leaving while sobbing is not the safest way to start a journey. I thus slip off early in the morning before the house wakes, as saying farewell the night before softens the blow of leaving. Despite feeling the ambivalences of leaving, these moments also heal, thawing painful experiences frozen since early childhood or another time in one's life.

Anticipation and leaving, like traveling, are human dynamics because they reflect the image of the Earthroamer. The theological depth of traveling is revealed in these dynamics long before we decide to travel. Every twist and turn away from home becomes an opportunity to learn something about ourselves, about this world we live in, and about God. Learning and discovering are additional reasons why abandoning travel plans is a good idea once traveling. Without opening oneself to discovery and even some purging, meeting God anew and being present to one's experiences remain elusive.

Born into Anticipation and Leaving

Anticipation—the expectation that something will happen—is an action, a practice, and an emotion, one that we first experience in the womb awaiting the moment of birth. There is a looking-forward-to element in our biology. As infants, that looking-forward-to action and emotion continued around our needs for food and also around basic bodily actions such as bowel movements and sleep. When we were hungry, cried, and anticipated food, it most often appeared, as if we miraculously produced it ourselves. Our parents or caregivers would leave the room, and we anticipated their return. Initially, anxiety entered our little bodies when mom, dad, or a caregiver left our sight, but we soon learned to anticipate their return. Of course, we also learned that we cannot omnipotently control the return of someone or the outcome of an event. Sometimes we dreaded what we anticipated, such as when we remembered the bad dream we had—possibly a night terror—and we anticipated the dream would return the next night. Bullies at school or gym class awakened similar apprehension and fear for we knew what was coming. Traveling takes us emotionally to times preverbal, when our experiences were intimately felt but we lacked the language and the mental capacities to process what was happening. We travel not only to places but also to relationships and communities, past, present, and future. It is both a grace and a burden that our emotional lives do not know time. What happened long ago feels immediate. Herein too lies the healing qualities of traveling as we unconsciously or mindfully revisit events of the past and now, as adults, rewrite the story of our lives.

Developmental psychology has helped us see that when a child feels secure, the child soon exhibits autonomy, initiative, and industry, psychological traits travelers rely on.[3] We see *autonomy* in the toddler who wobbles down the grocery aisle exploring the world on her own with a loving parent lurking close by. There was distance even then, for being too close interrupts the exploration. We know that autonomy looks different across families and cultures, and each of us practices autonomy unlike anyone else. Likewise, individualism, which follows autonomy, looks different in the West compared to individualism in communal cultures. Collectivist cultures emphasize the needs, desires, and goals of the group as a whole over those of each individual but do not deny one being an individual. Discovering the shades of autonomy is reason enough to travel.

Initiative, in turn, is often intrusive. The young child crosses boundaries for she is unaware of those boundaries' existence in the first place. Toddlers intrude physically but also by sound, for they are often loud. Being intrusive, going to and into places, is not morally laden, as exploitation would be. A traveler, even the most conscientious, is always entering. We enter foreign lands, homes, other spaces, even lives; we compare and find the other or ourselves lacking, a dynamic we'll tie to returning home. We take photographs without asking; we offer authoritative opinions, not always acknowledging our own parochial nature and the complex systems we've entered. At best, we intrude with tenderness and care, but we still intrude, interfere, and interrupt, just as toddlers do.

As we approached school age, autonomy and initiative were joined by *industry*. Industry speaks to the young child who makes things happen, whether through imaginative storytelling, building and breaking things, or playing games. It speaks of a sense of confidence that resists moments of feeling inferior. Industry indicates a child is ready for school and a future with knowledge as the child now does things alongside and with others. Such industry should not be confused with the neoliberal industrial complex, where persons are commodified or financially exploited to produce goods or services without cost-of-living compensation. Young and old rely on their sense of industry to scratch their travel itch and to sustain them while traveling.

Travelers also know moments when a deep sense of personal security and well-being eludes them, and feelings of insecurity, doubt, and shame set in. In such moments, home feels distant, even if they have not left yet. We anticipate according to what life taught us about expectation, and as such, it is a form of knowledge about something yet to happen. Anticipating holds some uncertainty.

Embracing Anticipation's Emptiness

To be human is to anticipate. Psychology helps us understand some of the motivations behind our leaving. Philosophy, in turn, informs a deeper understanding of anticipation and time, as these concepts are interconnected.[4] Of interest to us is how we understand an experience such as anticipation that bridges the past, the present, and the future. Anticipation is a form of knowledge about something that is yet to happen, a precognition, similar to experiencing *voorpret*. As a precognition, however, anticipation is, at best, "empty." Whatever we know about the future is incomplete. When we bracket our own judgments as we enter an experience, literally putting them aside, an experience has the potential to inform and surprise. Knowing our thoughts of anticipation are incomplete; we have to go beyond what we expect to experience fully. We have to cross a horizon, of sorts, and allow new experiences to surprise us.[5] This horizon is internal to the experience but also indicates infinite possibility beyond what is essential to that experience. We anticipate best not only when we seek to discover the essence of a relationship, a community, a locale, or an experience but also when we open ourselves to possibility. Neither the internal nor the external horizon will be discovered if we anticipate with judgment and "know" exactly what will happen. Instead, we nurture an attitude often called *protention*—having an openness to the future, reaching out into the future, living toward—rather than practicing *retention*, which ties us to a known past. Protention and anticipation that are not foreclosed go hand in hand. Empty anticipation is a present moment that has a past to be reckoned with, but it holds on to an open future. Travelers practice protention when they plan, with some detail, their next travel, only to change or even discard the plans once traveling. Protention embraces a form of open uncertainty, knowing that what we think might happen may not occur. We do so without the excessive anxiety not knowing brings.

Anticipation is different from expectation.[6] Anticipation, present in all peoples, has more "thickness" when compared to expectation. When we expect rain, we take an umbrella "just in case," even though we know from past experience that weather forecasts are often inaccurate. The present action is linked to the past. Anticipation is when we smell the rain and quickly close all windows and doors. The present action is linked to the future and even breaks with the past. Anticipation has a thrusting-forward feeling that propels focused action that expectation lacks. Though we are looking forward to something and we have a specific goal in mind, anticipation's paradoxical nature implies we are doing so with some indeterminateness.[7] It is not always clear what we anticipate, whether we anticipate positive or dreadful experiences.

Whether it is a baby waiting to be born, a toddler exploring grocery aisles, or a child heading off to school, explicit and implicit goals are behind the action of anticipation. As such, anticipation shares elements with hope, the dynamic of waiting for something to unfold, expecting an intervention of sorts. Both anticipation and hope trust that something will happen but have different starting points. Anticipation differs from hope in that we have more say in whatever we anticipate. We have a good idea what awaits us when we anticipate due to anticipation's protention. When we hope, however, we indicate our lack of control of both our current situation and the one to unfold. We anticipate travel, for which we prepared ourselves, by being open to new experiences. When we hope, something most often needs to change in our here and now, and we do not know if even a semblance of what we hope for will unfold in the future—we hope and trust it does. Hope, due to its liminal nature, is thus "riskier" than anticipation, as hope carries more uncertainty. "All tourists [and travelers] travel with a sense of hope," writes travel scholar Pearce Phillips. "They are hopeful in the teleological sense that they imagine themselves at a brighter point in their imminent future. Some may simply want to meet their hedonistic holiday goals while others may possibly seek to learn more about the world or create new identities. Further still, a select few may be seeking to support or even change the relationships in which they are involved."[8]

It is no surprise that anticipation, as a basic human trait, is an embodied experience. When an infant notices food, either as a breast or as a bowl, the infant stops crying. Our bodies prepare us for resolution of some kind. We experience this often when we listen to music, read a book, or watch a movie, and tears begin to trickle. The body's anticipation also appears when the smell of a food or perfume transports us to distant times, places, and relationships. The body's release or yearning for resolution drives our desire to travel when the body has been overworked, overstressed, and underappreciated. The tourist and resort industries know this dynamic too, fueling the illusion that extravagance and consumption will restore body and soul. Taylor and Jordan, trying to escape *karoshi*, discovered that their island rejuvenation lasted but a few days once back at the office as familiar rhythms resumed. Since our bodies anticipate, we are propelled toward an open, empty future.

We can, of course, also close ourselves off to the emptiness of anticipation and add layers of protection to our beings. A man had a business meeting in an African capital and added a safari visit to his trip. Before leaving, he anticipated poverty, social decay, politics tinged by corruption, slow internet speeds, and high crime rates. Having spoken with colleagues, looked over maps and travel brochures, and researched the location, he felt secure in that knowledge. Upon arrival, he was not surprised by what he saw, only by the depth of poverty, as precognition shapes experience. Sitting in the back of a large SUV, he saw barefoot children begging in the streets, a security presence, persons digging trenches by hand when a backhoe can do the same work much faster, and tenfold powerlines from a single electricity pole siphoning power to modest tin shacks. The businessman wondered why the kids were not in school and commented to his driver. With a tinge of guilt, he wondered where their next meal will come from. He knew he would eat a splendid meal at the hotel. There was little to no surprise. What was anticipated was what he experienced. Our businessman, agile when reflecting on world markets and opportunities, failed to recognize the deep sense of community that surrounded him and the solidarity, resiliency, and entrepreneurial ingenuity of the people, with their many curbside shops and industries. Creating jobs is paramount, and thus digging by hand, which gives employment to many, is weighed against digging a trench by backhoe. He did not see the relational ties that counter the radical individualism of the West. The community had a different understanding of time and seemed in no rush. They embraced controlled chaos.

Like many Westerners mixing business and pleasure, he is unclear about the legacy of colonization. The easy smiles that persons wear as they walk under billboards promoting a Chinese bank go unnoticed. There is touch, friendship, and eye contact; marginalized persons in need of care seem to have assigned guardians. Walking the city, he is invited into a small restaurant, but he remembers how some tourists have their credit cards cloned, and the invitation for hospitality evaporates. He takes many photographs to show his children back home how privileged they are. Since darkness equates to danger to our business man, he avoids the plazas and city at night and is fearful of dark-skinned persons. Invited to church on Sunday, he passes on the offer, for a three-plus-hour service without air-conditioning feels too much. Our business traveler soothes his conscience by buying some memorabilia at the hotel kiosk to support the local economy. He has something from Africa to hand his family. Opting for habitual certainty, he forecloses enriching experiences. Travelers who anticipate know better than to fall into false securities as they seek the next horizon to cross.

Anticipating is fraught with danger and definitely risky but not in the way our safety-obsessed culture conceives risk. Anticipation does not ask us to embrace danger. It is risky as it asks us to be vulnerable enough to open ourselves to new experiences, even ones imagined. When we bracket our experience, when we do not allow assumptions to guide our experience, and we suspend our judgment, we are more likely to discover the significance of an event or relationship. We anticipate our travel, therefore, with the awareness that, yes, we are pulled forward, but we are pulled toward a future that remains mostly unknown and empty.

Anticipating God in Time

Part of the emptiness of an unknown future is the Earthroamer who awaits us. We travel with the yearning to meet God in the stranger, in the city, in a remote location, on the road, in nature, even at the airport. As philosophy helps us understand human experience, theology too informs our understanding of anticipation if we approach it from below, from the perspective of the disenfranchised, the marginalized. Meeting God as our deepest longing is a central focus of theology. German theologian Karl Rahner reminds us that when we anticipate, we long to meet God as Divine Mystery.[9] In seeking the mystery of God, we acknowledge that our own subjectivity is complex and mysterious too, as no biology or psychology or philosophy fully describes being human. We know our knowledge is incomplete as God remains a mystery to us. Rahner calls this form of incomplete knowledge or preapprehension *Vorgriff*.[10] *Vorgriff*, like anticipation, is an imaginative act, though we may not be aware that this longing to discover the mysterious God is driving our actions. It is only when we encounter the Divine Mystery that we transcend our environmental limitations and the horizons of our own knowledge and claim not only our ignorance but also our transcendence, that we reflect the image of God.

As we anticipate God, we trust our experience and the people and places traveling takes us to. God awaiting us is part of our desire for meaning and knowing why the world is there or where we best fit in this world. We reflect on the significance of specific encounters as we open ourselves to the reality of God. In doing so, we transcend our own horizons, those limits of knowledge and self. As we anticipate then, we also anticipate transcending the spatial-temporal confinements of our embodied existence. We seek to go beyond what we know, especially about the Earthroamer. Anticipation as *Vorgriff* is a tacit awareness of becoming more aware of the universal in the particular. We experience this in visceral ways that touch our whole being. This is a *positive* infinity as *Vorgriff* seeks to find the infinite in the finite and quickens our desire for exploration and travel.[11] Preapprehending God is not only at the center of ourselves as spiritual beings—it informs our traveling.

We have already noted two aspects of anticipation that inform our exploration of time: anticipation's openness to the future and our longing to discover God as mystery and ourselves as transcendent. A third element that informs our exploration of time is our experience of time itself as past, present, and future blending together in anticipation. As any traveler will attest, we never have enough time to travel. Rather, the traveler has a complex relationship with time. When we anticipate, the future transforms the present and often rewrites the past. We have already said that time can be understood as *chronos* (moments determined by a clock) and as *kairos* (moments that are pregnant with possibility, that demand decision and action). This twofold understanding of time is deepened by seeing time in terms of *futurum* and *adventus*, as another German theologian, Jürgen Moltmann, argues.[12] *Futurum* means what will be; *adventus* means what is coming.

Futurum speaks to a future determined by a past and a present. As such, we know what to expect. *Futurum* (from the Latin *fui*, "I was," and the Greek *physis*, "becoming"; "things in process") speaks to our calendars and our futures unfolding. It is predictable, and the future can be extrapolated from the past. We will designate a day of departure, prominently displayed on our calendars. Some of us even create a large printable calendar posted on the refrigerator door and with a red sharpie cross out days as we anticipate leaving home. Our *futurum*—when seen as calendar entries—stretches out weeks and months and years. We count the days, for living a *futurum* life excites, but it rarely instills vitality as it offers no hope.

Adventus (Latin correlate for Greek *parousia*; "arrival," "entrance"), in turn, literally means the arrival or introduction of an "other." It is God breaking into our reality and creating something new. *Adventus* embraces the mystery waiting to be revealed and foresees a future that is different. *Adventus* does not exclusively point to Jesus and his divine purposes but also speaks to how God is revealed to us, daily and in diverse ways. *Adventus* moments redefine our past, as all who had a conversion experience can attest. They also determine our present experience and our future. Since the Earthroamer became part of our history, we are called to peace, justice, love, and reconciliation.

Addressing the wrongs of the past and present is a mandate. Jesus's life and death indicate such an *adventus* moment. When we experience an *adventus* moment—and they seem to be fleeting moments only—elements of God's reign manifest in our midst. We are surprised, we experience awe and wonder, the hair on our arms stands on end, and we discover ourselves and the other anew.

To travel is to find ourselves in the best possible position to experience *adventus* moments. Our day-to-day living is so *futurum*-oriented that the *adventus* moments seeking to surprise us daily go unnoticed. The traveler, whose imagination is based on *adventus*, is not satisfied with a future that will unfold but instead longs for those moments when the limits of our understanding or being are transcended. *Adventus* moments reveal God ahead of us, God around us, God in front of us, and God within us. They leave us as travelers with a new understanding of our role, function, and position in this world. Travelers also know that those moments do not last forever, so we seek every opportunity to experience another *adventus* moment. As we anticipate, we play with time, always in shortage, open ourselves to a known-yet-foreign future, and expect to find the Earthroamer along the way.

Anticipation as a Scriptural Subplot

When we read Scripture through the lens of anticipation, the Judeo-Christian narrative provides image upon image of our spiritual ancestors anticipating. An exhaustive list is not needed to show that being on the move is often associated with anticipation. The first time we meet anticipation is when Adam looked at all God had created and anticipated finding a partner; he discovered none. Adam was lucky, as his anticipation did not remain empty, even though he lost a rib as Eve was formed. Next, we find the couple anticipating God's coming to the garden after they ate the forbidden fruit. This time, anticipation was laden with guilt and shame as they went into hiding in fear and self-loathing. Meeting God, they are informed that they will be expelled from the Garden and told to expect enmity with the snake, pain in childbirth, and a land that will give up its food only with much labor. How this anticipation must have weighed on Adam and Eve! Moreover, they are told to expect death and a return to dust. We meet their sons Cain and Abel, with Cain anticipating God will find favor with his offering, and when that does not happen, Cain kills Abel in rage. Grace, however, remained. Cain, expecting that he will be killed for the murder, is told to anticipate being a wanderer and that he will remain unharmed by others. We've only covered the first four chapters of Genesis and we see that anticipation is a thread in all the narratives.

The Genesis narratives are archetypal in nature in that they portray deep truths about human nature and also about how we experience God. We discover that we were created to anticipate. Adam's anticipating a partner reminds us that our deepest longing is for mutual recognition and belonging. Eve, taking a forbidden fruit, did not do so to defy God. Rather, she took the fruit to address her longing to be transformed and be like God, to have knowledge. What failed her was the deceitful promise that eating something instills God's wisdom. Eve teaches us that anticipation and the desire to be transformed go hand in hand. As Adam and Eve leave the garden, their anticipation is tinged with fear and dread. We romanticize anticipation at our own peril. Cain, in turn, teaches us that anticipation can awaken feelings of disappointment and rejection, emotions powerful enough to guide our actions in thoughtless ways.

Scanning the lens of anticipation over other Scripture passages reveals similar patterns of anticipation that fuel hope, restoration, disappointment, and fear. Noah, for example, anticipated rain, while others, who did not anticipate being punished for their sins, thought he was crazy and ignored him at great personal cost—their lives. Later, Noah let a dove out, which returned with an olive leaf, and all in the ark anticipated receding waters and a new beginning. As God entered into covenants with Noah and Abraham, anticipation took on new meaning as it became fused with blessing and promise. As Noah's family grew, they anticipated they could reach the heavens by building a city with a tall tower. Never did they anticipate that God would bring confusion by introducing different languages, and the building stopped. Israel's history furthers this theme of anticipation as the people of God held on to God's promise of a land that would be theirs. Imagine the disappointment when Israel saw a whole generation die as they marched in the desert. The prophets, who anticipated God's reign of pathos, also anticipated the Messiah as the personification of that reign. This later surprised an expecting Mary, who did not anticipate being pregnant.

Turn the lens to Jesus and his followers, and the presence of anticipation continues. Going to the temple, Jesus thought his parents knew he was visiting his Father's house, only to discover they did not—they left without him, thinking he was among the other children in their traveling group. When Jesus went to a wedding, he anticipated wine only to receive water instead. Turning the water into wine became his first miracle. Later Jesus expected a temple space to reflect holiness and was met by merchants selling offerings. He became angry and toppled their tables. Jesus anticipated his disciples would understand his teachings; most often they did not, though many Samaritans did. Jesus thought that healing on the Sabbath would be welcomed, but it was questioned. The disciples did not expect to feed five thousand with five small loaves and two small fish. No one anticipated Jesus would walk on water. The Jews, anticipating a Messiah that would restore an earthly kingdom, had to make sense of a Jesus who died on a cross, a happening foretold by Jesus but disbelieved by even those closest to him. No one anticipated Jesus to be raised from the dead, even though he did tell his followers.

Many more narratives in Scripture carry the thread of anticipation—we just have to read with an open mind. Since we looked at the very first pages of Scripture, a brief look at the very last chapters of the Bible, also through the lens of anticipation, revisits the themes we first encounter in the narratives of Genesis. The apostle John's visions in the book of Revelation too are filled with anticipation, where it is colored by eschatology, matters of the end of time. The last chapters (Revelation 20–22) anticipate a time when Satan and the evildoers will be locked away for a thousand years. We are still waiting. Maybe we need to be disappointed. We are told next that Satan and the evildoers will be punished and are promised a new heaven and a new earth. We are still waiting. The God who hovered over darkness in Genesis is now the Alpha and the Omega, wiping away tears of mourning; there is no more leave-taking. People, we read, will never thirst again. We can expect a city that shines as crystal, with nations and rulers living in peace in the presence of God. We are still waiting. The last chapter of Scripture anticipates Eden restored and envisions a river flowing from God's throne into neighborhoods, watering lands and nurturing lives (see chapter 5). The Lamb, Jesus, then says, "Look, I am coming soon" (Revelation 22:7). Christians, for two thousand years, have lived in anticipation and with waiting.

This cursory look at anticipation as a lens through which to read Scripture reveals that we are indeed anticipatory creatures. The stories of Scripture follow the themes of openness, hope, and transformation we identified as central to the practice of anticipating. Anticipation and the experiences of love, loss, and shame, as revealed in Scriptures' archetypal stories, may be the only experiences we all share. Consistently, anticipation creates "thin," liminal experiences where the Earthroamer touches humanity, moments that change the destinies of persons, peoples, and the earth. Anticipating traveling is no different. Whereas the narratives of Scripture anticipate a time in the future where all will be revealed, travelers know that there is a day coming, marked on the calendar, when one departs.

Leaving as Disloyalty

We read in Scripture that one should leave the house of one's parents, find a partner, and become one. Leaving is just as difficult as finding a partner and becoming and remaining one. Poets, artists, philosophers, psychologists, and dreamers have contemplated leaving for millennia. Psychologies have been built around leaving and separating, as these actions reveal our earliest emotional ties. These ties are woven into relationships beyond mom and dad to include "allomothers," all the caregivers that acted in a parental manner toward us.[13] Allomothers are a comfort for travelers, as we inevitably need caregivers along the way.

How infants separate from their caregivers is well documented, even if we do not always understand why infants and toddlers separate in the ways they do. All infants *hatch*, seen when an infant crawls or walks away from mothers, fathers, and allomothers and then looks back as if to say, "Please keep an eye on me, for I am afraid to venture too far."[14] The infant, maybe eight months old, cannot separate without experiencing anxiety when encountering strangers. The hatchling then begins *practicing* (separation), something that comes naturally as the infant continues to crawl, climb, and walk. The distance away from caregivers increases with an infant's growing confidence. Seeing an adult run after a toddler who is off on her own journey brings frustration and laughter to the adult and bystanders alike. Practicing, in turn, leads to *rapprochement*, a time when the child ventures further away from her caregivers as new experiences are more attractive. The toddler experiences rapprochement as an emotional (self) and relational crisis in which hesitation and mood swings are common. Rapprochement indicates that the child keeps mental images of her loved ones alive in her mind, a process called object constancy, which sets in around age two. Through the process of internalization, caregivers live as mental images inside the child, allowing the caregivers to be out of sight but not out of mind. Rapprochement allows the child to be away from the familiar, finding inner comfort and safety. Through hatching, practicing, and rapprochement, the building blocks are laid not only for how we are likely to attach to others but also for our future traveling. Should traveling be new to our experience, following a toddler who hatches, practices, and returns in rapprochement offers a model to venture further to foreign experiences. As we anticipate our travel, we "remember" our initial venturing out in infancy, a memory that is coded into our bodies, as it predates language.

Our emotional ties and ways of leaving were formed in a weblike context that includes all our experiences in childhood. Patriarchal worlds masculinize leaving, as we find in Scripture's call for a man to leave his father's house. Travelers know gender is not an indicator of seeking adventure and journeying. Other times leaving is judged morally, seeing the person who travels like a hatchling—not venturing too far from home—as lesser than the person who pursues the ends of the earth. Traveling cannot be measured by distance, as we have said, but is measured in terms of attitude, imagination, and transformation. Power relations of gender, race, class, identity, and access to health care and education all shape our relational attachments, including the one we have with ourselves. Power judges someone "clingy" or "distant" or "secure" and "mature." We should remember that our emotional ties are socially constructed and not biologically determined, and they do not describe ethical living. *Just Traveling* resists families and systems that see gender and identity, race and class, and different abilities in narrow ways. Since every traveler has been taught much about exploring, we expect a flood of different emotions as anticipating and leaving begin to morph into one.

The leaving inherent in traveling invites the formation and transformation of healthy, reciprocal emotional relationships with self and others, including the relationships we have internalized and our relationship with nature. When our emotional ties reflect love, care, trust, and autonomy, they do present themselves differently compared to emotional ties that feel unsafe, absent, anxious, insecure, strained, avoidant, or distant, or ties that are colored by conflict or fused with reactivity. Though our emotional ties are strong, they are not impervious to transformation. We learn about our emotional ties, how they remind us of our internalized relationships, and how they determine our leaving. It is to be expected that approaching the moment of leaving awakens protest in some part of a person, possibly manifesting as irritability, despair, detachment, stubbornness, and anger. Travelers find themselves crying when watching a movie or listening to music or are strangely envious or suddenly bogged down by hesitation and ambivalence. Some sob. Others know misunderstandings and the fights they get into with those they love before leaving (and upon their return). Another part of a traveler's person, however, can be high on excitement and almost manic, ready for the unknown that awaits. We are complex beings, always becoming more ourselves as we separate from others while deepening our relationships to them.

Leaving—separating—is difficult for many reasons, some of which we've identified already. Facing physical and emotional distance from others is challenging, as our internalized relationships guard against experiences of abandonment. Psychoanalyst Jessica Benjamin reminds us that "separation includes giving up the narcissistic fantasy of omnipotence—either as perfect oneness or self-sufficiency."[15] When we leave, we acknowledge that we are separate from others and that we do not have control over what is coming. We admit that we have various needs that our current experience or lives are not satisfying. By traveling, we give ourselves a chance to become humbler, more empathetic, and also more grateful. For our emotional-spiritual selves, however, leaving is disloyalty despite the transformation that awaits us. We travel but also seek ways to make amends. As the moment of leaving approaches, questions that portray our ambivalence flood our minds: *Maybe I should not go . . . ? Can I leave those I love and travel to distant communities and destinations and enjoy new experiences without them? What if my loved ones need me while I'm away? How do I respond to my dad/mom/whoever it is that will say spending money on travel is a waste? Will I be faithful to my partner on the journey and will my partner faithfully wait for me and receive me back upon my return? What are we saying as a family when we tell others we're about to go on a trip? How will my partner and I or our family negotiate the different expectations we have for the trip ahead? How does one travel if it implies a growing carbon footprint? In what ways does my traveling support the capitalist system that drives a gap between the rich and poor—the very system I seek to dismantle?* There is no traveling without disloyalty and ambivalence, unsettling experiences that easily spill over into a relational, spiritual, or religious crisis. We are told to honor our mothers and fathers and to love our neighbors. *Where does loving oneself by traveling fit in?*

Our early emotional ties, because they are biologically driven and socially constructed, are also fluid, and we do grow out of them. It is grace that the *voorpret* we experience and the *wanderlust* we have are so strong that we overcome the ambivalences of leaving. Whether our attachment orientations and internalized relationships carry anxiety or not, we travel—though those of us with anxious or ambivalent internal relationships inevitably experience higher levels of anxiety when we leave.

We do not have to reenact our childhood travails when traveling, as we learn in life to regulate our emotional experience. Three practices assist pending travelers in regulating their emotional experience. First, telling about the upcoming travel prepares us and those around us. Every time the future is explored in conversation, our emotional life and relationships are adjusting to what is coming, the moment of leaving. Second, celebrating the fact that we, irrespective of age, are emotionally attached to a variety of persons and things can help letting go. Travelers anticipating travel will carefully wash their favorite travel jacket, or they may purchase a new one after much research. They weatherproof their boots or make sure their backpack has no seams that are coming apart. They ensure that their favorite shirt or dress or scarf is packed, for wearing it brings comfort. Travelers know to bring something of special significance along, even a little stuffed kangaroo that has been well loved. Since staying connected to loved ones and friends is important, travelers may set up a blog or other social media to communicate with friends and family from the road. Those off the beaten path may have a satellite tracker pinging their location to loved ones and, in case of need, first responders. Reading books, such as Robert Pirsig's *Zen and the Art of Motorcycle Maintenance*, Anne Lamott's *Traveling Mercies: Some Thoughts on Faith*, or John Stilgoe's *Outside Lies Magic: Regaining History and Awareness in Everyday Places*, to name a select few, awaken a necessary traveling attitude.[16] And third, we can ease the burden of leaving through rituals that support endings and beginnings. Functioning at the emotional and symbolic levels of our lives, rituals are powerful, as they do not always require words. They are powerful enough to contain emotion and overcome irrational (and rational) disloyalty. Sharing a special meal before departure or even a prayer or liturgy prepares all for leaving. Some travelers start something new, like a puzzle or home project, that will only be completed once they are back from their travels.

To Go Empty

Any reflection on travel has to address anticipation. Anticipation gives itself to us in ways similar to expecting the reign of God, in ways already here, but not yet. It is colored by hope, the pregnant wait for an *adventus* moment. It knows faithfulness, for looking forward to an experience always delivers some of its promise in the here and now. Anticipation is always focused on the present even as it entertains the future. Those who run too fast into the future, who do not allow anticipation to affect them, may miss their present experiences altogether. They may also forget that anticipation links the past, present, and future as a gift to be explored. There is a sufficiency in anticipating. It is enough until the day of departure arrives.

By anticipating we rehearse what is to come. We enjoy what is offered by the practice even if we know anticipation is but a hint and a promise.[17] Anticipation helps us answer questions every traveler receives from others who journey differently in this world. *Why do you want to go there? What are you traveling for?*

Why return to . . . , weren't you there two years ago? Are you going to [this city] too, you are in the country already? Where are you traveling to next? You must be spending much money on travel. . . . I heard your destination is not very safe right now. These questions and comments are not easy to respond to. We travel for many reasons, including to heal moments when time was foreshortened and experiences foreclosed, moments of crisis and trauma that touch every person. We travel to feed our *wanderlust,* our curiosity and interests, to be vulnerable and to feel lost and disoriented. We travel to experience the joy and pleasure of anticipation, as we discussed in this chapter. We travel for histories and present longings that cannot be met by any other means and do so in our individualized ways.

Anticipation is transformational just as what we foresee remains vivid. It is the start of one's travel weeks or months or even years prior to departure. *Voorpret* does not seem to care when the traveling will occur, only that the day of leaving is coming. Anticipating a brief getaway is just as life-giving as planning a cross-town trip to enjoy an ethnic restaurant and new museum exhibit that are making headlines. *Voorpret* also does not care about distance, as if going on an international trip is "better" than heading to the neighboring community or state or province. We best anticipate with an open mind bracketing our expectations, as foreclosing one's travel is missing an opportunity for transformation. Anticipation seeks the horizon, the limits of our experience, and awaits liminal moments. In doing so, we reflect the Earthroamer, who moved across voids, in gardens, across deserts, and toward, with, and in the people of God. We go empty, or, as we'll explore in the next chapter, we surrender to "the road."

"The future was plump with promise," Maya Angelou writes three days before her travel companion, Guy, was severely injured in "a brutally delicious experience" that came in the form of a car crash.[18] Guy survived the crash, and Angelou's traveling shoes were about to get dirty. But first she had to cry. A friend, Efua Sutherland, tells Angelou, "Sister, you need a sister friend because you need to weep, and you need someone to watch you while you weep."[19] Angelou receives this graceful invitation and cries, letting go of pent-up anticipations and disappointments—about Guy, about her past. She had to travel from America, via Egypt and on to Ghana to do so. As Angelou drains herself of tears, she remembers her grandmother telling her as a child that "the more you cry, the less you'll pee, and peeing is more important." Angelou chooses to be disloyal to her grandmother. She surrenders her past, and her future opens through tears. The gifts of anticipation and leaving—of distant communities and destinations—always surprise. *Voorpret* remains true to its meaning.

Chapter 2

SURRENDERING

Longing for Letting Go and Renewal

> [Surrendering's] ultimate direction is the discovery of one's identity, one's sense of self, one's sense of wholeness, even one's sense of unity with other living beings.
>
> —Emmanuel Ghent, "Masochism, Submission, Surrender: Masochism as a Perversion of Surrender"

> All of us have breakdowns around particular affect/event experiences where we "lose it" so to speak; all of us preserve, unwittingly, antiquated habits of mind that no longer functionally serve our purposes.
>
> —Barbara Pizer, "When the Crunch Is a (K)not: A Crimp in Relational Dialogue"

We travel to liberate ourselves from the many (k)nots that keep us entangled. (K)nots, therapist Barbara Pizer writes, are those early childhood experiences later "encoded and elaborated in procedural memory as habits of mind that exist outside awareness and are as much taken for granted as the very breath of life."[1] They are those habits, predispositions, and behaviors we bring into our relationship with ourselves and others that ensure superficiality, distance, or stuckness. (K)nots appear in our persons but also in our partnerships, in our parenting, and in our professions and deny authenticity, intimacy, and even promotion. We are engaged but not really involved. We dance an old dance, one we do not like but one we need not think about either. The private, communal, societal, and political ways we get ourselves tied up, however, seldom exceed our creativity in getting ourselves freed. There are paths to freedom, and traveling is one of them. It is not uncommon for travel to stem from some sort of impasse that has left us frustrated, bored, stagnated, angry, or needy. We travel to feed our curiosity, excitement, and creativity, recognizing that our here and now has become inadequate. Whatever the reason for our traveling, our (k)nots remain with us until they begin to untie and let go. The possibility of traveling as a path to freedom remains.

Since traveling holds a cluster of promises to alter the ways we experience ourselves, persons, communities, places, nature, and God, traveling includes seeking release from that which binds. This desire to get away and to get released describes a kind of surrender, something the self actively pursues. We do not want to remain frustrated or needy for too long. To be clear, the surrendering we explore is not the surrendering to a power in defeat. The West has viewed and continues to view surrender negatively, for the West prefers the master and only glances at those without power—the poorly compensated worker, the marginalized, and the victimized. In the context of power relations, we do not surrender but submit, or we become sadistic, definite (k)nots to untie.

Crossing a fast-flowing stream in the Colorado Rockies on a motorcycle, the invitation to surrender and be transformed came to me. My partner Barry and I were following the Continental Divide Trail (CDT) from Roosville (Montana) on the Canadian border to Antelope Wells (New Mexico) on the border with Mexico. The 2,500-mile trail traverses through awe-inspiring mountains and furnace-like deserts. Through-hikers of the CDT definitely got bitten by the travel bug more than once or their (k)nots are plentiful! By the time we reached the stream, we already had a long day after the route became impassible, and we ended up following an extended detour. Despite having lost the trail a number of times and having minor tip overs, I had not yet surrendered and felt "I can do this." Sometimes surrendering only comes about through some adversity.

Late in the day, we ran into the stream fed by recent snowmelt. I became worried that we might have to backtrack. Impatience when traveling rarely adds positive value to the experience, and here it was no different. While Barry assessed the stream, the overconfidence I had in my motorcycle and my abilities led me to plunge in without walking the stream first. *Surely I can power my way through the stream*, I thought. Two-thirds of the way across, however, I sunk into a hole with water rushing at seat height. My motorcycle stalled and sucked in water as I struggled to keep it from toppling. Visions of floating downstream flashed before me. Barry, clearly angry at my carelessness, slowly got off his motorcycle, took off his riding boots and pants, and waded in to push me to the other side. Walking back, he discovered a path across, half as deep and with no holes.

We now found ourselves on a remote dirt road. My motorcycle needed an oil change, but calling in help was futile, as our cell phones showed no reception. No trucks or vehicles came our way, and the silence between us reflected our situation. I was wrestling with my foolish impatience, and Barry was wrestling with me. Part of me wanted Barry to lash out at me so that the foolish me could be punished. I broke the silence by saying that I would walk along the dirt road until I found a farm where I could ask for help. Barry waited by the motorcycles to flag down vehicles that risked the stream and were heading in our direction. As Barry placed his camping chair and opened a book, I started walking.

This walk became an opportunity for introspection. It was a walk of surrender and to freedom, one that invited me to accept some truth about the situation and me. I recognized our situation was serious but not hopeless. At some point, people will pass this road and summon help or I'll find a farm. Where my thoughts led to next had more possibility. I got us into this mess due to a strong tendency to naturally claim control to escape frustration and to be recognized. I cannot hold my impatience very well. I did not think the CDT or the stream sought to teach me a lesson, but I did think that I could learn from my situation. I confessed to God and the high desert that riding the CDT was primarily ego inspired, the checking off of a bucket-list item, something to boast about. I contemplated how to apologize to Barry while honoring our amazing adventure. Walking in sturdy off-road motorcycling boots not made for hiking added to the experience. As I walked and talked to God, memories of walking alone into the unknown flooded my awareness: I am six and walking alone into school for the first time, emotion flooding my body; I am ten and riding my red bicycle to a new school after moving towns and the same feelings appear. I feel alone, confused, and small. One of my (k)nots is a forced stubborn independence birthed when I needed emotional support in childhood and it never came. Who plunged into the stream, me or my distant, independent father in me? As I walked, tears started to flow over my cheeks in cathartic release as I felt the loosening of these past (k)nots. I slowed my pace and meditated on a phrase I sang in a choir many years earlier—"Ek kyk op na die berge . . . ?"—to quiet my inner voice's blaming and shaming.[2] This felt better than asking God straight out for help, a weak attempt at manipulating God. How does the Earthroamer save us from our own poor judgment and foolishness? Does the Earthroamer turn water into oil?

A few miles down the dirt road, I saw a farmhouse in the distance. Despite warnings not to enter, I opened the gate, which triggered a silent alarm, and within a minute I saw a man coming toward me in a truck. When I told the man—Maurice was his name—what had happened, he just smiled, saying that motorcyclists underestimate the stream often. We returned to his barn where he got tractor oil, the only oil he could offer me. Sometimes travelers have no choice. My lesson about my controlling ego continued. Control always seems to keep something from happening. With Barry and Maurice watching, I got ready for the oil change, something I have done many times, but this time I forget to open the oil filler cap. Oil trickled out at a snail's pace as darkness approached. And then I remembered the cap. Red-faced, I told my onlookers things will move faster now. With the oil replaced, we left Maurice—who refused payment for the oil, a muddy oil mix—and continued down the dirt road to find a campsite.

To surrender, then, is to yield to an experience and to get the part of the self that wants to control and be reactive out of the way. It is to discover and integrate elements of one's self, to let go, to be present, and to see the world with new eyes. Traveling and surrendering are intimate partners.

Traveling's *T-Third* and the Self's Longing to Surrender

The call to surrender falls strangely on Western ears. Emmanuel "Mannie" Ghent, a renowned psychoanalyst who practiced in New York City, focused our attention on surrender, which he differentiated from submission and masochism.[3] Though Ghent wrote about surrendering in 1990, we only now are plumbing the depths of his wisdom. Surrendering is a therapeutic concept that describes the experiences of both the therapist and the client, but it is not to be confused by contexts of imbalanced power or conflict where the word also has meaning. It is of special significance in the exploration of the travel experience. Western culture sees surrendering as defeatist, as a form of kenotic emptying in the face of danger or power. In the East, the Japanese know *amae*, a concept much like surrender. It is a valued virtue. *Amae* speaks to a sense of dependence, tolerance, and understanding. *Ameuru* as a verb reflects the meaning of surrender as explored here. It leads to *satori*, enlightenment, and *kenshō*, seeing the nature of one's being with an awareness of how one fits into the whole.[4] *Amae* reminds us that independence does not guarantee freedom and, conversely, that dependency also fosters freedom.

The surrendering the self longs for describes an action seeking to show the authentic self, one that resists being compliant, especially to outside emotional demands. We learned this compliance early in childhood as we read the emotional states of our caregivers. Intuitively we knew that to receive care and to avoid neglect and even abuse, we needed to tread lightly around adults who were loud in their anger, silent in their distance, or caught in their addiction. Environments where love is conditional and threats lurk discourage authenticity and fuel reactivity and compliance in a child. These early adaptations become a permanent way of engaging the world. Psychology talks about this self as a false self, passive to impingements. The false self is a caretaker self, protecting us when the environment does not care for or nurture us or is deemed dangerous. We also recognize an authentic or true self, that part of every person that remains creative, spontaneous, and playful. It is the true self that longs to be recognized and be known. The true self resists the (k)nots in our lives.

Masochistically, the false self seeks ways to force the environment to recognize us, especially in relationships where others have power over us, but such force pushes the authentic self into hiding. The false self is not open to surrendering's promise for a new beginning. The true self knows better than to fall for false adaptations, which it experiences as unreal.

"Surrender implies the freedom from any intent to control or coerce. . . . Some form of surrender, that is, acceptance of the 'bad' feelings—helplessness or destructiveness being chief among them—is required to really reclaim the position of responsible agency," writes therapist Jessica Benjamin.[5] As we surrender, emotion floods our bodies. Anxiety, fear, and dread increase, so also feelings of abandonment, for the caretaking function of the false self is relinquished in the self's search for cohesiveness. The false self is creative in keeping us away from the unsettled state of discovering our true selves. The caretaking self thus often puts us in the position of the one being done to, as someone else's power or influence sadistically flows over us. As we masochistically become compliant yet again, there is the hint of recognition but not the kind sought by the authentic self. In our relationships, we do not seek to relate false self to false self, though we do that almost all the time. Rather, we long to be discovered and recognized by our partners; to be seen in such a way that we need not hide or present our naked vulnerability only in darkness, under sheets, or for a fleeting moment. As we surrender, we discover ourselves and long for intimacy with others, the Divine, and nature (see chapter 3).

Saying that we pursue traveling to surrender to an experience, or to participate in an experience, needs further comment. We actually surrender to a *third* entity that is neither ourselves nor travel per se. Counseling often talks about "the therapeutic third" as the facilitative environment created by and between a therapist and a client. This third is more than what the therapist and client individually or mutually create, for the sum of the third is more than its parts. As I have explored elsewhere, technology creates a similar transformative third, much like a therapeutic third.[6] I would argue that traveling creates a transformative third as well. We understand this third, which we can call *the traveling third* or *t-third* for short, as the total experience created when our travels put us in moments and situations. We cannot will the *t-third*; many contributors—our own self, persons, communities, places, times, moments, and more—play a role in creating this transformative space. The *t-third* as a space is both conscious and unconscious and real and imagined, which makes it odd, even mysterious, for we know it is something, but describing it runs into the limits of our imaginations and language. Suffice to say that traveling creates experiences much like a therapist or a guru cocreates a transformative *third*, guiding and recommending and inviting the self to deeper integration.

We best surrender *in the presence of* others rather than *to* someone or something. Here traveling assists us as we surrender in the presence of strangers, persons we may never meet again, or to nature. The anonymity aids the process of letting go. Surrendering is always *a voluntary activity*, since it is devoid of power relations that control or command. Resisting surrendering, which we do, is thus different from resisting relationships of power, control, and abuse. We cannot force surrendering by sheer will as if we omnipotently create a *t-third*. Traveling provides the facilitative conditions needed to make surrendering a possibility. The process of surrendering includes *powerful emotions*, as we've indicated. The range is wide, from feelings of dread, anxiety, guilt, and fear of dying to having clarity or experiencing relief and ecstasy. Some seek to withdraw and be alone; others want to be with people. Feeling fully alive, *we are in the moment* as our past and present blends into an integrated experience with our future. Whatever bothers us loses its sharp focus. There is release as we are more at peace with who we are and what is happening, and a *sense of unity with all around us* emerges. "I" and "mine" become "we" and "ours." There is no "they," as we often call the other that we dislike or the persons or groups we do not know or understand. This is the experience of deep solidarity with others, nature, and the Divine the self longs for (see chapter 4). Once we have surrendered, *a sense of wholeness*, contentment deeper than happiness, sets in. We travel to discover the undefended lovable self we are. We surrender indifferent to power, which generally prohibits the integration we long for due to power's exploitative nature and purposes. As we surrender, a deep *sense of acceptance* sets in. We accept not only who we are but also what we are experiencing. We accept the *t-third* as a gift that seeks to nurture a new self, one hidden for many years.

To surrender, then, has little if anything to do with power, domination, or seductive excitement. It is not hoisting a white flag. Rather, it is the self's search for freedom beyond the fulfillment of any parental, societal, political, or divinely sanctioned expectation. Surrendering, which needs broad description to counter the negative connotations it has in the West, is a practice with transformative potential that liberates and expands the self. Embracing surrender asks us to let go of defensive barriers the false self brings, such as *catastrophizing* (making things worse than they are); *minimalizing* (diminishing a feeling, thought, or experience); *selective abstraction* (not taking the whole picture into view but rather focusing on one minor detail); *dichotomous thinking* (either-or thinking); *drawing invalid inferences and conclusions* (not reading a situation correctly); *gross generalization* (assuming one's experience is every person's experience); *predicting behaviors* (calling an experience before it happened); *magical thinking* (attempting to manipulate through ritual and thought); and other distortions, evasions, and diversions of our reality.[7] Surrendering resists falling back on anger, which briefly will empower a person when one feels vulnerable, exposed, or disempowered. Traveling promises a self that is more open, also open to transcendence, truer to itself, more connected to others, and less defensive—and a self that feels alive.

When we travel, *invitations* to surrender abound. For some, the moment of surrender will arrive in the act of embracing and receiving the unknown; for others, it will be in a meditative moment. It might be walking a pilgrimage path, alone in the presence of others, or hiking unexplored backwoods. Some travel to discover art, to paint and create. For others yet, it will come in more active forms, letting things happen or in a moment of (orgasmic) ecstasy. It might be overcoming deep fear in an act of faith and plunging off a bridge, dangling on a bungee cord, or it might be buying a ticket to fly to a foreign city with no plan about what to do upon arrival. Some trek deep into the Amazon forest to experience ayahuasca with an Indigenous shaman. Distant communities and places are attractive. Surrendering seems to come more easily when we are removed from the ever-present glare of screens and the overstimulation technology brings. Regardless, to surrender is liberative and open-ended. It seeks growth and expanding to deeper levels of wholeness, unity, and solidarity. Beyond liberation, surrendering is a form of resistance and counterresistance. We resist the compliant self, formed in less optimal environs, touched by neglect and conditional love, emotional or physical abuse, loss and shame. Some of us grow up in the shadow of parents whose own narcissism suffocated or adults whose partnering created anxious insecurity in the children around them. We also resist corporate worlds that provide a lifestyle but take one's life (see chapter 4).

We surrender and travel to achieve a higher self, to the false self's chagrin. Simply put, we travel to come into being. And since this is a gradual, spiral process, requiring a lifetime for most, we return to the practice of traveling, each time with the same hope that we will surrender a different part of ourselves. It is grace that we carry tacit knowledge that surrendering is life-giving. Surrendering is not about objective knowing but rather about subjective understanding. When a person cannot surrender due to strong defenses, it is not only sad; it is tragic. Whereas there was a time those defenses served the person well, even helping them survive, the false self does not recognize that the defenses are no longer needed. Defenses serve us well in the short term only. We travel in faith that the *t-third* will provide, even if the process of surrendering is unsettling.[8] Travel demands that one show up as a whole person, able to be responsive to what lies ahead.

Travel with All One's Heart, Soul, and Strength

Whereas surrendering is a concept of growing interest in psychology, a theology of surrender that supports the practice is not easily found. Jesus's life of submission is not the first thought that comes to mind, for one recognizes elements of submission and even self-harm in his life as he starved his body for forty days and walked to the death he knew awaited him. In our violent world, new understandings of atonement are sought.[9] Sacrifice and power play key roles in the Judeo-Christian tradition that resonate with the West's understanding of *surrender* and its relationship with honor and shame. *Just Traveling* uses surrender in a different way, denying tradition's exclusive use of the word and practice. Turning to theological reflection on the Trinity is not of significant help either, as the relationship between the Creator (Father), the Redeemer (Son), and the Sustainer (Spirit)—called perichoresis—poorly translates into human relational terms and does not correlate with surrendering as an emotional longing. To deepen our understanding of surrendering, we turn instead to three separate presentations of wisdom in Scripture. These areas of exploration describe aspects of a life of faith. The first is the Greatest Command(ment), the second is the apostle Paul's insight into how God's Spirit works in us, and the third is how we understand the nature of God's grace. These pieces of wisdom further illuminate how and why travelers surrender and often see those moments of surrender as pinnacle spiritual experiences.

First, *Just Traveling* recognizes that we are called by God to love God with our whole being. The *Shema*, an ancient Jewish prayer proclaiming God's sovereignty and describing faithful living, states: "Hear, O Israel: The Lord our God, the Lord is one. Love the Lord your God with all your heart and with all your soul and with all your strength" (Deuteronomy 6:4–5). As Jesus engaged experts in Jewish law, he recalled the *Shema* and added wisdom from Leviticus, "Love your neighbor as yourself" (Matthew 22:39; Leviticus 19:18).[10] Jesus radicalized the *Shema* by saying we need to love our enemies (Matthew 5:44). In a few sentences, Jesus gave us "The Greatest Commandment." The unity of the Earthroamer, that God is One, demands an integrated human response. For Near Eastern people, heart, soul, and strength represented the totality of being a person. They saw that the heart was the center of one's emotions, especially love. The soul represented the consciousness of a person, now infused with love. Love in this context is ambiguous, for it likely points to both religious and political loyalty (as Near Eastern covenant treaties show) and to filial love and affection, as well as erotic love (Genesis 34:8; Deuteronomy 21:11). The reference to strength indicates that it takes intentional effort and action to love God in a covenantal relationship. Jesus's broadening of the *Shema* turned what was emotional, religious, and political to be also personal, relational, and ethical. We are called to care (see chapter 4), a state of being that demands the integration of our various parts, especially our loving and our hating. Jesus states that when we love ourselves, our neighbors, and our enemies, the reign of God is near (Mark 12:34).

We see a relationship between travel and surrender in Luke's version of Jesus reflecting on the *Shema*. Here, Jesus continued his reflection on the *Shema* by telling the parable of the Good Samaritan (Luke 10). We learn of a Samaritan man, from a people despised by the Jews, who traveled between Jerusalem and Jericho. The winding road was known as the "Way of Blood" due to the ever-present danger of robbers. The parable tells that our traveler happened upon a fellow traveler, a Jewish man, recently beaten and robbed. The Samaritan man became personally and financially involved to ensure the unfortunate traveler was nurtured back to health. One may question why the Samaritan man traveled on a road he knew was dangerous. Or why, as a man of means, he did not bring along protection against the bandits. We do not know whether the Samaritan man experienced surrender, though traveling's *t-third* does foster such moments, but we recognize evidence of surrender in his care ethic and solidarity with a neighbor in need. The priest and the Levite, however, who also passed the man in need, chose not to help, though the reasons for their inaction are unclear. The parables, New Testament scholar Amy-Jill Levine reminds us, often leave us with more questions than answers.[11]

Surrendering is a total body response that brings deeper awareness of our connectedness with all around us as well as with the Divine. It compels new actions, ranging from slowing down, tending to what previously was avoided in anxiety, and seeking to make a difference in one's immediate environment, which, maybe only hours before, was foreign. When we surrender, we experience not only oneness with ourselves and others but also a moment of unity with the Creator, the Earthroamer. It is as if the moment of surrender manifests the reign of God in our midst, a fleeting reality. This happens in very visceral and concrete ways. Suddenly our relationships with God, neighbor, enemy, and nature are redefined. The *Shema* and Jesus's expansion of this ancient expression of faith mirror surrendering in that they anticipate an integrated person who is responsive to what life brings. It takes one's whole being to live according to God's guidelines, as it takes a whole person to surrender. When we experience a moment of surrender, we are in an optimal position to love God, self, and neighbor with all our heart and soul and strength. Such living demands an integrated, liberated person who lets go of past pain and conflicts and the grudges they bring. We reach mountaintops and deep valleys of spiritual and personal integration by allowing travel's *t-third* as a transformational moment to touch us.

The apostle Paul's vision of how the Spirit works in us offers a second area of faith that speaks to surrender. In his letter to the Galatians, in a section translators often call "The Fruit of the Spirit," Paul reflects on a person who lives by the Spirit of God: "What happens when we live God's way? God brings gifts into our lives, much the same way that fruit appears in an orchard—things like affection for others, exuberance about life, serenity. We develop a willingness to stick with things, a sense of compassion in the heart, and a conviction that a basic holiness permeates things and people. We find ourselves involved in loyal commitments, not needing to force our way in life, able to marshal and direct our energies wisely. Legalism is helpless in bringing this about; it only gets in the way."[12]

Paul draws on the prophets Isaiah and Joel, who foresaw a fruit-bearing Israel due to God's promises of restoration, which includes God's pouring out God's Spirit on all people.[13] What Paul also describes is a person who surrendered. The experience of surrender includes healing, restoring, removing that which depresses, and erasing and correcting the ways of the past. After moments of vulnerability, we often let go of what kept us tangled and (k)notted, and we grow less defensive. To surrender is to feel alive, to have the exuberant joy of emotional and spiritual highs as fear and anxiety are let go of. It is to grow in persistence. When we feel burdened and burned out, we skim from partner to partner, from employer to employer, and from exotic destination to exotic destination seeking excitement. Each passage, however, offers but short-lived excitement, and as a heaviness sets in again, we repeat the cycle of excitement. The surrendered person, keenly aware of being one with others, is compassionate (Latin *patí*, passion; *cum*, with) and acts proactively, for persons and all of nature are seen as sacred. Paul pictures relationships held together by love and loyalty, not power. Walking away from power that controls, abuses, and wounds is a first step toward surrender. The person with the Spirit, Paul states, is a good steward, able to manage time and one's abilities and resources well. His warning that legalism cannot bring forth the fruit of the Spirit mirrors our inability to will surrender. We look at a person who surrendered and see the fruit of the Spirit.

The fruit of the Spirit and surrender share some purposes, such as the building of character and deepening interpersonal relationships. Both rely on a liminal experience that transforms a person, nurtures a new identity, and finds persons in cooperative relationships. Psychology reminds us that high levels of cooperation lead to persons being more "socially tolerant, empathic, helpful, and compassionate."[14] Tests show that cooperative persons have higher levels of self-esteem and more secure identities, and they are more responsive and resourceful in their lives. Through the Spirit and after moments of surrender, we are vulnerable and open to perceive and experience reality beyond narrow personal beliefs, opinions, and needs. In our gendered world, women welcome the Spirit and moments of surrender more readily than men do. In her *My Brother's Keeper: What the Social Sciences Do (and Don't) Tell Us about Masculinity*, psychologist Mary Stewart Van Leeuwen describes an experiment of Christian college students whose perceptions of the fruit of the Spirit were assessed. Respondents were asked about words like "*strong, submissive, patient, emotional, kind, self-reliant, self-controlled, nurturant, aggressive, ambitious, joyful, modest, faithful, tough and loving*."[15] Distributed through the list were the fruit of the Spirit: love, joy, patience, kindness, faithfulness, gentleness, and self-control. While all the traits were mostly labeled as neutral (versus masculine or feminine), the male subjects labeled the fruit of the Spirit as feminine. As the fruit of the Spirit falls victim to gender bias, so too surrendering as a practice that invites a new sense of self, one that is caring and compassionate, is challenged. That many travelers ascribe an experience of surrender as the working of God's Spirit is a truth we should accept and not dispute, for God's grace nurtures growth and transformation.

Surrender and the Judeo-Christian tradition share *grace* as a dynamic that transforms, a third area of spirituality and theology that speaks to surrender. Grace is a "benevolent divine influence acting upon humanity to impart spiritual enrichment or purity, to inspire virtue, or to give strength to endure trial and resist temptation."[16] It speaks of divine favor or goodwill that leads to restoration and renewal. The word's etymological roots speak of favor, benevolence, and gratitude. For Christians, grace is often synonymous with God's mercy, salvation received through the crucifixion of Jesus, and receiving the sacraments. Grace, however, like most elements of the spiritual and religious life, is not monolithic. Dutch scholars of the nineteenth and twentieth centuries, following the reformer John Calvin, introduced an understanding of grace important to the practice of surrender. They differentiated between the *common grace* God offers and the *special or particular grace* of Christ as Savior.[17] These theologians came to common grace as they wrestled with a question revealing their Calvinistic roots: *How can there be good in this world if humanity is sinful and depraved and our reality broken?*

Gratia communis—common grace—was first named by Calvin as a grace that covers all of humanity. Common grace is named as the force behind the order one finds in all of creation; to explain why, despite humanity's brokenness, humankind does not destroy itself; to notice all people as spiritual beings; and to celebrate that all persons carry gifts that enrich not only their own lives but the lives of others too. It identifies a benevolent, caring God sustaining all of creation. Louis Berkhof writes that common grace "is not redemptive and yet exhibits many traces of the true, the good, and the beautiful."[18] Likewise, Abraham Kuyper acknowledges that "One is struck by . . . the remarkable fact that, weighed against the doctrine of our depravity through sin, the unconverted world exceeds our expectations; and the church, weighed against the doctrine of the re-birth, disappoints our expectations."[19]

Just Traveling follows common grace. As the descriptor "common" indicates, this is grace that is universal to all humankind and not dependent on religious practices or beliefs in general, or on a relationship with Jesus in particular. It is the undeserved divine grace bestowed on all, breaking through every social category we have. The people of God readily claim God for themselves, though the prophet Joel is clear that the compassionate God gifts all people with the Spirit of God: "[God] is merciful and compassionate, very patient, full of faithful love, and ready to forgive. . . . I will repay you for the years that the cutting locust . . . have eaten. . . . After that I will pour out my spirit upon everyone; your sons and your daughters will prophesy, your old [folk] will dream dreams, and your young [adults] will see visions. In those days, I will also pour out my spirit on the male and female slaves" (Joel 2:13, 25, 28–29 CEB). The biblical witness insists that common grace is not uncommon or exceptional. Every person receives this gift from the

Earthroamer. No one controls God's grace as it flows in, to, and through all people.

We fault any theology that identifies *gratia communis* and then proceeds to build vast hierarchical doctrines around election (or those elected), providence, and the special grace some theologians identify. Common grace is redemptive and holistic and denies spiritual reductionism, which prides itself in saving the soul. The special revelation of God, it seems, easily triumphs over the general revelation of God. Kuyper rightly argued, however, that common grace has its own content, purposes, and scope and "cannot be placed in subservience to God's reconciling, redeeming and electing work in His covenant of grace," Jesus Christ.[20] Special grace argues that the restoration common grace offers is insufficient. We disagree with such a narrow view, which seems threatened by the possibility that God extends grace to all. Common grace is the Earthroamer's gift to all of creation, travelers included.

Surrender and common grace mutually inform each other. Both speak of a universal longing to be restored, to let go, and in that longing, every person is related to every other person sharing the same longing. The concepts mediate personal experience and transformation. Both surrender and common grace identify humanity as inherently spiritual before they engage in religious practices. Persons have faith in the possibility to be restored and in their ability to make a difference in this world whether they have specific religious and doctrinal beliefs or not. Furthermore, both common grace and surrender see God as the benevolent Earthroamer, who journeys with the traveler and sustains humanity. Common grace reminds those who surrender that behind their experience is the Divine, sometimes obviously present in tangible ways, other times hidden and mysterious, barely recognizable. It is grace that we find the Earthroamer in moments of surrender. Common grace grants every person the possibility to be restored, to not be defined by past, present, or future. Surrender, in turn, challenges common grace to accept that all persons long for experiences to let go and be restored and renewed. Surrender brings individual focus to common grace's wide horizon.

The traveler embracing common grace can expect moments of surrender as manifestations of the benevolent Earthroamer, who promises restoration. Without God's common grace that prepares persons, communities, and destinations to be welcoming and caring, traveling's transformative potential will be greatly diminished. With common grace as a backdrop, we can say with confidence that traveling is not only an invitation to experience the good and beauty that is in this world, in people, in cities, in nature, but it's also an invitation to experience God. Philosopher Martin Buber in his *I and Thou* writes, "The Thou meets me through grace—it is not found by seeking . . . The Thou meets me. But I step into direct relationship with it. Hence the relationship means being chosen and choosing. . . . All real living is meeting. . . . The relation to the Thou is direct. No system of ideas, no foreknowledge, and no fancy intervene between I and Thou. The memory itself is transformed, as it plunges out of its isolation into the unity of the whole. . . . All real living is meeting."[21]

Meeting another person in authentic ways takes an undivided wholeness, an authentic self that recognizes the authentic other self. This meeting, which we'll discuss in the next chapter, requires the immediacy and focused attention of surrender. To travel is to open oneself for transformation. It is also an invitation to witness the destructiveness in and to the world, the inequalities that keep people oppressed. This witnessing calls on the traveler to make a difference in this world, to travel with choice (see chapter 4). Common grace beckons us to join the Earthroamer, who restores and protects, the God who enlivens people with the Spirit.

Thinking theologically about surrender, especially as it informs the theology and spirituality of traveling, asks what it means to live authentically with all of one's heart, one's soul, and one's strength. We long for the same unity we notice in the Earthroamer, the One God. Surrender also reckons with the work of God's Spirit in all people, a work that reflects the grace God bestows on all creation. Traveling is deeply spiritual but also transformational because God's unity calls forth our wholeness and integration. We experience spiritual moments because God's Spirit is alive in us and God's grace is found everywhere. The etymological relationship between words such as healing, whole, and holy becomes the lived experience of the traveler.

Travel and the Pursuit of Transformation

The kernel of surrendering and the self's longing for transformation are laid down in infancy. Parents and caregivers, through their care and provision, altered the self-experience of their infants before a blanket or soft toy was discovered as a soothing object. Later in infancy, we used beloved blankets, soft toys, and pacifiers or thumbs to moderate our emotional experience. As adults, we use traveling to be transformed and to moderate our internal worlds. The above objects are paradoxical, as they are discovered by the child as a first possession, though they were there all along. These objects, sometimes called transitional objects, are part of the holding environment that facilitates the development of a child. As such, the holding environment is much bigger than the parent-child relationship. Rarely will a parent leave the house with their child without taking the object along, for it will soothe anxiety and sadness and diminish fear when the child encounters newness. As the child holds the teddy bear, the teddy bear holds the child. The teddy bear, then, like the mother and the environment, becomes a *transformational* object, "experientially identified by the infant with processes that alter self experience."[22] A child is thus held in at least three ways: by the mother or parental figure, by the environment as a whole, and by the child's transitional object. This holding facilitates transformation of inner and outer worlds, including the integration of the physical—the emotional and cognitive—and the environmental into a young child's being. We assume that this environment is loving and caring and consistent, but most often it is lacking.

Whereas children have a *possession* recognized by all caregivers as important, adults *pursue experiences* that will transform their sense of self. We do so, psychoanalyst Christopher Bollas writes, "in order to *surrender* to an experience as a medium that alters the self, where the subject-as-supplicant now feels himself to be the recipient of enviro-somatic caring, identified with the metamorphoses of the self."[23] We have failed to see in adults the longing to alter self-experience, which takes many forms. It is common for this pursuit to revolve around food, drink, drugs, sex, religion, service to others, employment, gambling, speed, aggression, and risk. What characterizes humankind is that we pursue experiences that transform our self-experiences. Our imaginations and selves are creative in pursuing these transformative experiences. Some seek positive experiences; others seek what they hope would be a positive transformation of the self, but it ends up as another (k)not. Our pursuit of transformational experiences acknowledges some deficiency or an awareness that something is missing in our lives. The travel industry banks on our desire to be transformed through exotic travel experiences and erotic brochures. Travelers, however, know that there is no easy path to spiritual enrichment and personal transformation.

When we travel, we are held by something bigger than we are—the *t-third* created by the Earthroamer's grace. Like a child, we are vulnerable, but we are also kept in such a way that transformation is possible. Our inner worlds are adjusted. This does not go without some emotional or even physical reaction, which can be unsettling. As such, the pursuit of traveling is an expression of hope and confidence in the future. We hope, each in our unique ways, that the (k)nots will be loosened and that we'll be renewed. We anticipate moments of surrender. Travel's promise of transformation, through moments of surrender, is liminal, uncanny, and spiritual. We travel because we have early memories of being transformed, first by a mother or a parental figure, then by the environment and those objects we endowed with special powers. The way our caregivers appeared with food when we needed it, or changed our diapers when we cried out in discomfort, or left the light on when we were afraid of going to bed—so travel awaits us, ready to receive those open to transformation. We long to discover the truth about ourselves. We think less of who we are not and more of who we are becoming. Surrendering takes us one step closer to this hoped-for self, a person we will have to reintroduce to others upon our return (see chapter 5). This moment is an experience a traveler does not want to miss.

Chapter 3

MEETING

Overcoming Distance and Celebrating Difference

> The more one is able to leave one's cultural home, the more easily is one able to judge it, and the whole world as well, with the spiritual detachment and generosity necessary for true vision. The more easily, too, does one assess oneself and alien cultures with the same combination of intimacy and distance.
>
> —Edward Said, *Orientalism*

As travelers, we visit places and step into communities where history denied relationships across difference and created gross inequalities. In his satirical piece "How to Write about Africa," Kenyan author Binyavanga Wainaina writes, "The Modern African is a fat man who steals and works in the visa office, refusing to give work permits to qualified Westerners who really care about Africa."[1] This man resists development and troubles the efforts of nongovernmental organizations. Or the modern African man may be an Oxford-educated intellectual who is now a corrupt politician and murderer. Wainaina advises authors to use words such as "darkness," "safari," and "sunset"; to write about "the starving African . . . how Africans have music and rhythm"; and to "describe, in detail, naked breasts." Wainaina is exposing the ways the West, in history books and travel narratives and brochures, represented and still represents the peoples and regions it conquered or the beaches, savannas, or forests Westerners visit. Academics often refer to this form of representation as Orientalism, brought to our attention by the influential Palestinian cultural critic Edward Said and characterized by seeing foreign peoples and places as exotic, romantic, also dangerous, and by grouping diverse peoples under single categories.[2] The fifty-four countries in Africa become "Africa": persons from Mexico, Central America, and South America are "Hispanics" or "Latinos"; persons from Asia with epicanthic folds in their eyes are "Chinese." Orientalism disregards the vast differences between peoples, cultures, and regions. It describes peoples in stereotypical ways while it performatively shapes engagement and expectations along Eurocentric lines.

There is an intricate love-hate relationship between especially the West (the Occident) and the East (the Orient). A similar relationship, however, exists between the North and the South. That we do not have familiar alternatives to describe the North (septentrional/septentrion) and the South (meridional/midi) and use terms such as "Europe" and "the Global South" reveal our biases. A traveler steps into this world of complex and ambivalent relationships where cultures are continuously and imaginatively created and renewed. At the same time, every person has a certain *location* that cannot be described by either the coordinates on a map or the insights cultural competency brings.[3] Location speaks to personal and communal identities shaped in a cauldron where race, gender, sexual orientation, class, age, ability, economy, generational position, institutional affiliation, and geopolitical locale become a matter of alchemy, ever mixing in ambiguous ways. As such, there are no "pure" identities, except possibly for the few remaining uncontacted peoples in remote rain forests.

Inaccurate and judgmental depictions by travelers are as old as writing itself. The Greek writer Herodotus traveled to Egypt and North Africa in the fifth century BCE and misunderstood Egyptian customs. In the second century CE, European travelers described Africa as exotic and dangerous. By the fifteenth century, we see the rise of European *imperialism*, where one nation or empire occupies and governs remote territories. When empires send military forces and workers to dominate countries economically, culturally, and politically while suppressing the existing local forms of societal living, imperialism evolves into *colonialism*. Colonizers typically see the colonized as inferior, ascribing to them that which colonizers repress about themselves—being primitive, hostile, and sexual—and regard the colonized as open to exploitation and even eradication. Colonialism never tires of new peoples and lands to exploit. Peoples exploited by European empires through the twentieth century are now dealing with the presence of China and the United States in their countries or are governed by the International Monetary Fund or the World Bank. *Postcolonialism* exposes the othering of those who do not mirror the colonizer or dominant culture. It refers to the time after the end of colonialism when a country received independence from the colonizer *and* to the time after the era of colonialism started, for colonialism does not end when a nation receives independence. *Decolonization*, the active work of undoing colonization by challenging the perceptions and institutions formed by colonization, further describes the world a traveler steps into. Though these constructs may feel abstract and academic, travelers know that billions of lives experience these forces as an intimate enemy in a daily wrestle. Powerful forces—including scientific classification, evolutionary theory, and Christianity's missionary zeal—are at work shaping our understanding of this world and our personal and communal identities.[4]

"The rhetoric of empire," with its Eurocentric metaphors of otherness, superiority, inferiority, and sensual and sexual desire for the other, remains.[5] Colonizers coded people according to appearance and customs, historical alliances or disagreements, and access to natural resources and wealth, and then associated themselves with one group or tribe and empowered that group to dominate the other groups or tribes in the region or country.[6] With social and cultural unity destroyed, domination received less resistance.[7] Every traveler finds themselves either closer to historical or contemporary colonizers or closer to the colonized, past, present, and future. Though travelers seek to meet persons in the liminal space, that space is contested and easily becomes congested. Travelers of color and with different abilities experience their otherness in ways white travelers may find difficult to imagine.

The Challenge to Meet the Other

A few experiences reveal both my location and the difficulty connecting with others when traveling. Heading to Key West, I stop at a convenience store to get gas and to hydrate. A tattooed man with a cane and cigarette in hand walks over and looks at my motorcycle. "I had many motorcycles," he tells me. I inquire about makes and models to get the conversation going. He tells me that he was a Hells Angels captain and rode various Harley-Davidson motorcycles "all over the South." My mind races as I imagine his past activities and I try to decipher his numerous tattoos. I listen but do not hear. His grandson, maybe five or six, joins us and shoots me with a nerf gun in the form of an assault rifle. As the foam bullets bounce off me, the man tells me he takes his grandchildren for breakfast on Saturdays. I notice two older children in a minivan, playing games on phones. The boy reloads his gun and, as the firing resumes, asks me if I have been pulled over by the police. I answer a quick no and continue my conversation with his grandfather. Getting bored with things, the boy tells his grandpa he is hungry, and we say goodbye. I never ask their names but gladly receive their blessing for traveling mercies.

Another time I am in Armenia. I lectured to leaders from neighboring countries and anticipate my return to the United States the next day. I share a ride with three young men newly converted to Christianity. They are in the area once a week to play soccer with teenagers. One speaks broken English, inquires how to study in the United States, and translates for me as I try to communicate with the other young men and our driver. With Mount Ararat in the distance and darkness approaching, we race toward Yerevan, the capital. We pass gray, smoke-stained apartment buildings dating back to the Soviet era alongside extravagant estates behind high walls with security guards. The young men have a heated discussion with our driver in Russian. The driver speeds up, almost touching the vehicles in front of us before he overtakes one. I reach again for the seat belt to be reminded that there isn't one. Suddenly the English speaker touches my shoulder and says, "We told the driver he is going to hell for not believing in Jesus. He is a Muslim. Maybe he will listen to you." I know this is not the place to admit I have long-unanswered questions about evangelization and about heaven and hell. I say that maybe Jesus's death is enough for all people and that having this kind of conversation without a relationship often fails, but my reply does not satisfy. I'm relieved when we roll into Yerevan and the soccer players are dropped off. I feel a need to apologize to the driver, but without language, I'm stuck. We ride the last minutes to my hotel in silence.

The road to Yerevan awakened a memory of an earlier experience in Bangladesh. Parbatipur in the country's northwest is not an easy town to visit. Close to the border with Nepal and Bhutan, it is an overnight train ride from the capital, Dhaka. Parbatipur is in one of the poorest regions in the world. Only 0.3 percent of the population is Christian. I am speaking to leaders of an organization sending doctors and nurses, engineers, teachers, and business leaders to the neediest people in the Middle East and Asia. The local hospital's chaplains, on hearing that I was a chaplain in the past, invited me to shadow them as they made home visits to rural communities nestled between miles and miles of rice fields. Despite some apprehensions, I am riding pillion on a motorcycle as we dodge traffic and poor road conditions. We enter the home of Shriti, a woman known to the chaplain, who translates for us. I learn that Shriti's husband is a construction worker in Qatar and sends money home. The small two-room mud home with dirt floors is sparsely furnished and clean. A toddler and his sister peek at me from the second room. We play peekaboo. Shriti tells me life, at times, is difficult. Her husband does not regularly send money to support her and their three young children. Since she converted to Christianity, her Muslim family refuses to help her. The chaplains at the hospital help her financially when the need is dire. Shriti is proud that her oldest is in school, an experience she never had. As she hands me a cup of tea to drink, I touch her gaunt fingers and notice her thin gold rings. I comment that I cannot imagine what her life is like. I mention the pride I witness in her, also her losses in not having her husband with her and not being able to connect with her family. I thank her for sharing about her life. As our visit comes to an end, she brings me a Bible and asks if I would read and pray for her, which I do. I cannot read Bangla but paraphrase the book of Joel for us. I remind her that God is a compassionate God, full of love and slow to anger; that God sees the damage the locusts of life cause; and that God is a God who restores. With hands held in prayer-like fashion, we part. I am struck by Shriti's faith and resilience, her hope that her children will have a different future. The chaplain and I get on his motorcycle and continue down the bumpy road. As my time concludes, I leave some money for the hospital to continue to assist women like Shriti.

Another moment that speaks to the complexities around meeting and mutuality occurred in Zambia. I'm on sabbatical and teach for a month at a university. I have been there twice before and have some familiarity with the university and the fast-growing city. Due to a lack of resources at the university, I paid my own airfare and rent a guest house to give additional income to the school. The house comes with Maria, a domestic worker, and David, a gardener. I feel conflicted about employing Maria and David and consult with my hosts, who tell me that this work is Maria's and David's only source of income. They are grateful I came, as the guest house was empty the past weeks and Maria and David had no income. Hearing what laborers such as Maria and David are paid, I inquire if paying them more would be possible, but I'm asked not to do so as it will create tensions across campus if some workers are paid more than others. I am told that not all "employers" are able to pay what I am willing to pay. I remind myself that Africa knows community and togetherness in ways the West has forgotten. Also, I remind myself that class differences are a colonial legacy my hosts seem unlikely to address. I am a short-term guest and will not be able to change societal agreements such as the ones I am invited into. Feeling I need to do something, I tell Maria and David to work half-days, and I purchase additional groceries to share with their families. We often talk, and I learn much from David and Maria, about how HIV/AIDS has ravaged their families, and about their hopes and fears for their children. Both are entrepreneurial and sell used clothes for extra income. David is saving money to buy a minivan he will use as a taxi. Maria teaches me how to make *nshima*, a staple much like grits made from *mielie* meal. David educates me on the names of the abundant birdlife, geckos, and lizards. Still, we never narrow the racial, power, and class divides between us. Discovering that the guest house will remain open after I leave, I pay Maria and David two weeks' wages as a driver waits to take me to the airport. We shake hands as I leave, David and Maria each holding their right elbow with their left hand.

Sometimes we meet nature and not persons. It's June, and I am following "Old Route 66" from Chicago to Santa Monica. I packed my motorcycle with camera gear and, reaching Flagstaff after hundreds of pictures, I turn north toward the Grand Canyon. I have camped in the national park before, but I'm called back. I leave my Tusayan hotel early in the morning to chase the golden hour before sunrise. The cold air stuns my lungs and numbs my fingers on a clear day as I ride to Mather Point. I feel fully alive. There is little traffic entering the park at five o'clock, but as I reach Mather Point, there are many photographers setting up their tripods. The morning is brilliant. I struggle to concentrate on what I am doing. My fingers are cold and sensitive when I touch my gear. As the sun appears and the colored layers of the Grand Canyon glow in shades of red and white and gray, awe and wonder flow over me. I feel blessed, privileged, small, and insignificant. I cannot really fathom how big the Canyon is, but it feels beyond time and bigger than the ten or so miles across. I vow to use time better when I am back home. I wish my partner and daughters were here to experience vistas that resist description—there is a definite downside to solo traveling. I know my photographs will disappoint as a poor attempt at mirroring creation. I notice a haze layer in the distance, smog blown in from fires, cities, mines, and factories as far away as California and Mexico. I wonder if "clean" days exist. As six thirty approaches, I put away my gear and grab the sandwich, juice, and apple I bought for breakfast. I walk the rim trail toward Yavapai Point, find a spot away from the growing crowds, and try to take in as much of the Grand Canyon as possible. People come and go. Few stay long. I recall reading in park surveys that persons are in the park for five to seven hours but look at the Canyon only seventeen minutes. It is ten o'clock, and I am hot in my riding gear. I head to the Visitor Center to watch the orientation film in the air-conditioned room. Tomorrow I will head to Seligman to continue along Route 66.

Meeting another person or nature beyond brief encounters and finding ways to exchange lives and experience awe and wonder are difficult, but such encounters deepen the rewards of traveling. *How can we meet others in English if it is the language of historic and modern colonizers?* "Every colonized people in whose soul an inferiority complex has been created by the death and burial of its local originality finds itself face to face with the language of the civilizing nation; that is, with the culture of the mother country," writes Martinican psychiatrist Frantz Fanon.[8] If language is an immediate barrier, a traveler needs relational skills and capacities to meet persons across differences. This chapter explores possible ways to think about and enter into relationships with the people we meet along the way.

Growing in Relational Intelligence and Cultural Humility

"To know how to sustain an ongoing human relationship means to know what it means to be an 'I' and a 'me,' to know that I am an 'other' to you and that, likewise, you are an 'I' to yourself and an 'other' to me," writes Turkish American philosopher Seyla Benhabib.[9] As two "others" meet, *relational intelligence*, "the art of being in authentic, ethical relationships with persons and easy-going, balanced relationships with things," is needed.[10] Since the capacity to relate to another is never a given, it is an art requiring a lifelong apprenticeship. We first learned this art when we looked into the eyes of caring loved ones and our mirror neurons, governed by cognitive and neurological processes we do not fully understand, got triggered. Our loved ones smiled, and we smiled too, but we did not mirror everything we saw.[11] This mutual recognition between mother and infant, which carries into all relationships, is central to human existence.[12] For an infant and traveler alike, mutual recognition creates possibility for affirming, validating, acknowledging, accepting, understanding, empathizing, taking in, tolerating, appreciating, offering goodness, and seeing that does not stare, ogle, or gawk. Mutual recognition was fleeting in the travel moments I shared. Power disparities and differences of gender, race, and class, with their imbedded privileges, are seemingly unbridgeable chasms.[13]

The mutuality we seek is different from the "*rules of recognition*" instituted by imperialism and maintained in a postcolonial world.[14] Rules of recognition are those ingrained customs, ways of thinking, and ways of relating introduced by a colonizing force that function at a preconscious level. We act in certain ways without thinking, for it is in the air we breathe, but when called to mindfulness, we do acknowledge our biases and recognize that the world is burdened by injustices and inequalities, or we remain blind to those biases. These rules typically ensure that sociopolitical and socioeconomic divides remain. They kept Maria, David, and me in customary, hierarchical, and questionable positions. They kept my host community from being willing to address how they perceive their togetherness. Since both mutual recognition and rules of recognition shape personal identities, any traveler has an invitation to grow out of ways of relating that reflect power over mutuality, exploitation over care, and stereotypes over deep authenticity. Research, however, indicates that knowing about cultural differences, even sociopolitical realities, may not be enough to save us from rules of recognition. What a traveler needs is *cultural humility*, a core aspect of relational intelligence.

Physicians Melanie Tervalon and Jann Murray-García explore how medical care is provided across diverse cultures. They cite studies showing how ethnicity persistently determines the care one receives, how some ethnic groups are perceived to experience less pain or discomfort. Their work on cultural humility helps travelers seeking ways to relate with authenticity across difference. Many persons are culturally competent—they know much about other races, religions, cultures, cuisine, and countries. Cultural competence is a "detached mastery of a theoretically finite body of knowledge" as "facts" about ethnic groups and others are studied, much as Orientalism does.[15] The problem with preconceived knowledge is that its "insights" and "wisdom" are filled with prejudiced notions about gender, race, and class, and about sexuality, nationality, age, different abilities, and other intersectional categories. Cultural humility does not rely on knowledge in books and online or on past personal experiences. Rather, it is built around three principles: a lifelong commitment to self-evaluation and self-critique, redressing power imbalances in the relationship, and developing mutually beneficial and nonpaternalistic partnerships (which includes systemic changes).

The knowledge and expertise we carry from our cultural competence and experiences create a false sense of security that will deny the authentic relationships we seek. Travelers risk missing visual, verbal, and nonverbal invitations for authenticity and human solidarity. Our past knowledge and experiences only serve us as travelers if we remain "flexible and humble enough *to say that [we] do not know when [we] truly do not know*."[16] Being humble enough to say, "Please help me understand . . ." opens the door for mutual recognition and negates rules of recognition. This requires mindfulness and reflexivity, traits few embody naturally. Rather, we have to learn not only how to uncover the biases we bring but also how to bracket our experiences and reactivity to keep them from impinging on our relationships.

We overcome the power imbalance between us and others when meet each other at eye level; when we show interest in someone's life, relationships, and society; when we are educated by someone else; and when we refrain from giving advice. We ask how to participate in what they are already doing, we acknowledge their competence and expertise, and we seek ways to collaborate on new joint projects. It is when we recognize others and are responsive to the power imbalances in our relationships, which change with every barrier of difference we have to cross, that mutuality is experienced. Together we identify and dismantle those systemic patterns and forces that keep persons in a power imbalance or deny people a flourishing life.

Travelers who embody cultural humility find themselves in positions where meeting others or communing with nature is possible. It is these moments that, if we are mindful, teach us much about ourselves and others. Meeting folks for a twenty-minute conversation at a convenience store is not the best locale to bridge divides, though it makes evident the differences that exist between people. A homestay for a few days is a better choice than peeking into someone's home through a door or window to "see how the locals live" and then returning to an air-conditioned room. Meeting someone is thus more than just being in the physical presence of that person or knowing about a community, a culture, or a destination. Meeting asks that we constantly overcome varied forms of distance and difference. It is discovering a person or community for who they are to themselves; it is listening to others and to what nature wants to tell us. Cultural humility is seen when we allow someone else to become our teacher. It is discovering someone else's hopes and dreams, customs and practices, likes and dislikes, as well as their successes and fears.

Gloria Anzaldúa, a Mexican American cultural theorist, identifies powerful metaphors that deepen our understanding of cultural humility. One metaphor is the importance of *nos/otras*, which speaks to the tendency of persons to separate themselves from others. *Nos* is all who are privileged and hold power; the *otras* is the other, the colonized group and all who are marginalized. "Nosotras gets divided into nos/otras, into an us/them division. The us/them dichotomy locks us into a who-is-more-oppressed dynamic. Internalized racism and internalized shame get played out."[17] Anzaldúa's use of Spanish, native words, and dialects—a practice called bilanguaging—forces the reader to travel, to be uncomfortable, to recognize the relationship between language and identity, to embrace cultural humility.[18] "Joined together, nos + otras holds the promise of healing: We contain the others, the others contain us. Significantly, nos/otras does not represent sameness; the differences among 'us' still exist, but they function dialogically, generating previously unrecognized commonalities and connections."[19] Traveling is the opportunity to experience nos/otras, to be shifted—transformed—toward mutual recognition.

We need practices to overcome the nos/otras divide, to overcome our blind spots and other prejudices, and to heal "*la herida abierta*," an open wound that is formed when borders and barriers between people cause harm, especially in the one who holds less power.[20] Anzaldúa invites us to a relational way of being called *conocimiento*. "*Tu camino de conocimiento* requires that you encounter your shadow side and confront what you've programmed yourself (and have been programmed by your cultures) to avoid (*desconocer*), to confront the traits and habits distorting how you see reality and inhibiting the full use of your *facultades*."[21] *Conocimiento*, a different way to speak of cultural humility, implies connection with others, portraying care and solidarity, being open and receptive especially to the spirituality of others, acting on new knowledge of the other gained. It is to resonate with someone else's hopes, dreams, and fears; allowing the people we meet to have voice and express what needs to be said and heard. *Conocimiento* demands "self-reflection, imagination, intuition, sensory experiences, rational thought, outward-directed action, and social-justice concerns."[22]

One additional Anzaldúan concept illuminates cultural humility. When we recognize deeper meaning in surface phenomena and find comfort in ambiguous spaces, we experience *nepantla*, being in-between physical and sociopolitical spaces, spaces where power manifests itself differently than in society. "Living in *nepantla*, the overlapping space between different perceptions and belief systems, you are aware of the changeability of racial, gender, sexual, and other categories rendering the conventional labelings obsolete."[23]

Nepantla is an undetermined place of constant transition in which difference disappears. As such, it is a liminal space. Here, assigned (binary) categories, such as man/woman, powerful/privileged, and rich/poor are questioned. Rather, our shared humanity is honored. Travelers, finding themselves in *nepantla*-like positions and facing the *nos/otras* divide, exhibit a new *conocimiento* through the ways they see and listen, through what they eat and read, and through how they are experiencing the touch of and by others.

Cultural humility reflects values such as mutual recognition, comprehension, respect, love, dialogue, and solidarity. Self-reflection—an act that counters willful ignorance—asks why a conversation could not happen or got derailed. It evaluates one's words and actions, of omission and commission. Self-reflection recognizes the complexity of personal and ethnic identities, of communities and national affiliations. Keeping a journal is helpful in this regard, and debriefing one's experiences offers transformational insights. Without a way of living that includes a lifelong commitment to self-evaluation and self-critique, meeting turns into an encounter. By traveling, we discover fellow humans beyond Orientalism and its children colonialism, racism, sexism, homophobia, and other -isms—forces that deny mutual recognition. Trust is built if rules of recognition are set aside. Travelers explore their own prejudices and stereotypes, they fault memory as being shaped by colonial and imperial powers, they recognize power imbalances and how those impact relationships, they expect colonial trauma in the lives of persons and communities, and they build the trust needed to be vulnerable and bridge across difference. At the center of all these actions, we find mindfulness and self-reflection, which determine not only how we look at persons and things but also how we listen, read, touch, and eat while traveling.

Core Practices in Meeting Others: Seeing, Listening, Reading, Touching, and Eating

Liminality, crossing over perimeters and borders, and passing through horizons are all inherent in traveling. We meet the other beyond our home or apartment; across neighborhood, county, and state lines; and past national borders. Moving between worlds and encountering margins that have been defined by sociopolitical forces masterful in marginalization and denying difference ask for a different way of traveling. These forces protect dichotomous views and values such as legal/illegal, national/alien, insider/outsider, sameness/otherness, object/subject, passivity/activity, settlement/displacement, and center/periphery. Crossing various borders exposes how closely one lives to the sociopolitical forces that protect separation, oppression, and injustice. Escaping these forces should not be confused with leaving one's home. Travelers function as intermediaries between cultures and their different experiences of reality, and they witness the search for wholeness in all. Having guideposts as one enters borderlands is helpful.

How and what one sees and looks at define much when traveling since "travel implicitly involves looking at, and looking relations with, peoples different from oneself," writes media professor Ann Kaplan. She warns that "[our] looking relations are never innocent. They are always determined by the cultural systems people traveling bring with them."[24] Kaplan highlights the eye, which has been called "the organ of tactility"—we touch and contemplate the world through primarily visual cues.[25] With passport in hand and a dialect on one's tongue, we are, as travelers, increasingly self-conscious about our national identity and we are more aware of gender, race, and class differences. Visually, we travel in at least four ways: We *look*—"determined by history, tradition, power hierarchies, politics, [and] economics";[26] we *gaze*—a kind of looking that is preoccupied with the self and its desires more so than being interested in the other, as when one person objectifies another;[27] or we *see*—making visual contact in such a way that it deepens mutual recognition. Anzaldúa refers to the gaze as a *mal de ojo*, an illness of the eye caused by staring (and also the evil eye), which she compares to *ojo de venado*, having the soft eyes of a deer.[28] Looking and gazing are exploitative and voyeuristic one-way interactions. When we look and gaze, we see our deepest desires or that which we despise in ourselves. The soul of the person we gaze at goes into hiding to protect itself. Neither looking nor gazing is open to self-criticism and mindfulness. Looking is passively guided by deeply ingrained patterns of relationship. Seeing, in comparison, anticipates mutuality. Whereas gazing is active in meeting desire—watching pornography, for example, which leads to sexual arousal or disgust or both—seeing is active in building trust and mutual recognition. Whereas looking and gazing objectify others, seeing honors the subjectivity of people and the sacredness of nature. When we see another, when deep recognition occurs, we experience the other as a Thou and not an It, as holy and not a thing.[29]

Those following the Earthroamer have a fourth way of using their eyes: they *witness*, a kind of seeing. To witness is to tell about an event or a person but also to observe, to notice. This second meaning is important as travelers notice how Orientalism (de)formed them and others; we notice conflicts and confrontations, dilemmas, distortions; we also witness hopes and fears, courage, and liberation. We are asked to leave our power and privilege behind. We learn to imaginatively see through the eyes of someone else, a challenge if we enter borderlands. Through witnessing, we discover our interconnectedness with humanity rather than focus on the differences we observe. We are implicated with the other through being a witness even if we find ourselves in a position where we are not asked or able to facilitate change. Witnessing is transformational when we inquire about what we see and so release persons and communities from their shame and self-doubt—what they are experiencing is not a figment of their imagination as the colonial masters said, it is real, and we are witnessing injustices and inequalities. Likewise, we witness their hopes and resilience, their joys and sense of community.

Travelers who approach persons and nature through a camera lens or phone screen risk turning seeing into looking, as images often maintain stereotypes and power imbalances along gender, race, class, ethnic, religious, and other intersectional lines that are not easily overcome. Even the language of "taking" a photo suggests mastery over, possessing, manipulating, and accumulating. When we look and gaze, we reflect the compromised conditions of our psyches and souls and society's inability to escape forms of postcolonialism. Seeing, witnessing, recognizing, contemplating, and self-reflecting undergird the ethic of traveling and honor the subjectivity of others and nature.

Listening is a second practice that facilitates meeting. It has been described as a lost art.[30] Listening is a difficult practice for persons coming from an outspoken culture such as the United States or coming from families where children were seen but not heard. Travelers best heed the warning of theologian Dietrich Bonhoeffer when he writes that "Christians . . . so often think they must always contribute something when they are in the company of others, that this is the one service they have to render. They forgot that listening is a greater service than speaking. Many people are looking for an ear that will listen. They do not find it among Christians, for these Christians are talking where they should be listening."[31] Listening serves the relationship we have with others. It has a simple measure: if we use more words than the person or group we listen to, we are not listening anymore.

Listening to someone else, especially across difference, requires the capacity to suspend one's own agenda. One must remain mindful of—bracket—defensive emotional reactivity, which difference stirs. The mindfulness is then extended to contain preconceived notions and expectations, of memory, desire, and judgment, additional bracketing. Listening remembers we exist for others. It follows a person—"tell me more"—rather than leads a person somewhere—"have you thought of . . ." It employs compassion—feeling with people. We respect the other person and come across as genuine and authentic, rather than being defensive or withdrawing. By talking about concrete things—the lives we live and what we are experiencing—we remain in the present. Escaping to the past or the future minimizes our ability to listen. These basic elements of listening are not easily embodied, but since listening is an art, practicing and improving our listening skills bring new possibilities.

Travelers listen to those they meet but also to music, podcasts, and news media as we enrich our experience. Many cultures are rich in storytelling and, combined with the creative spirit of musicians such as that found in griot traditions, provide stories and songs that seek to rewrite Orientalism and postcolonial histories. Making music is an ancient tradition, as even the Psalmist questioned singing in strange lands (Psalm 137:4). On playlists and streaming sites, music awaits discovery. Many an artist interrogates and probes and embodies *parrhesia*, an ancient Greek concept of speaking one's personal experience of truth.[32] Deeper than the West's understanding of free speech, *parrhesia* is a form of truth telling that seeks to break silences, chooses to be frank rather than to persuade, and accepts the risk of doing so in the face of power structures. Practicing *parrhesia* is always courageous and often deadly, as many a prophet will attest. Listening to songs that speak of poverty, of persons having no rights and being exploited, of women and children as victims of interpersonal violence, of the need for healthy sexual practices amid HIV/AIDS, and of corruption prepares a traveler to meet others along the roads they travel.[33] Storytellers and singers who practice *parrhesia* are modern shamans—the Sanskrit word for "shaman," *saman*, means song—and create new identities by mediating between the past, the present, and the future.

A third practice toward facilitating meeting in the *nos/otras* world explored by travelers is *reading*. We read to discover or deepen our knowledge of peoples, communities, and destinations. The relationship between newsprint and literary texts, digital news and social realities, is complex, as these arenas of life too are (de)formed by the powers this chapter names. Literary and other texts sustain, question, or rewrite persons and societies. The differences between colonialist propaganda and texts of prophetic resistance are obvious. Some travel brochures, narratives, and books do attempt to operate as if social worlds with histories burdened by colonization do not exist. They often hide behind images of sunsets, couples lounging by a pool, humor, and selective historical reflection, rarely engaging marginalized or silenced voices. In the texts they read or listen to, travelers question how power and ideology are presented; how persons, communities, and systems are portrayed or disrupted; they notice who and what is seen as unknown, displaced, distant, or dangerous; they are sensitive to the use of color—white, black, dark, red, and so forth; they query how history and heritage are understood, revisited, and rewritten. Travelers note whether Indigenous words are used; they are mindful of how time and place are described.

Texts, novels, and poems, like songs and instrumental music, introduce a place and its people to travelers and show how history is being (re)written. The creative spirit of contemporary writers is undermining colonial power.[34] Histories from the perspective of the colonized are being written.[35] So too classic literature is rewritten in ways that defy Orientalism and colonialism its powers.[36] Artists use mimicry, a recognized form of undoing colonial and other histories, to explore and describe the experience of the marginalized and the oppressed. They show the complex and ambivalent relationship between the colonizer, who traveled and conquered, and the colonized, who was subjugated and fought for an authentic identity.

A librarian can assist a traveler in finding books that play off each other.

Touch is a fourth practice that either builds trust or spirals into a traumatic moment. In a world where embodiment is a contentious reality, some travelers will shy away from touch, which is not possible. We know persons who crossed ethical boundaries past and present, guilty of invasive, penetrative, and other forms of abusive touch, which is immoral, illegal, and evil. Too many persons hold frozen moments, moments of abuse that reside deep in one's psyche, where they are best left undisturbed so as not to thaw.[37] Touch is always an ethical event. However, in the way Jesus invites a doubting Thomas to touch his wounds (John 20:27–28), touch confirms the existence of the other and informs meeting the other. We envision the unguarded and innocent moment of touch. It is the child who touches one's leg and looks up with hopeful eyes; it is the guide who holds one's hand while disembarking from the boat or bus; it is the waitress's hand touching ours as food is passed; it is the person on the bus whose hip touches ours; it is the person in the checkout line who backs into us; it is an arm that reaches out when it senses the person approaching has stumbled. These brief moments of touch remind us the other person is there, offering us an invitation to affirm that person, whether with a smile, a "thank-you," or a longer conversation.

The touch we explore is paradoxical in nature. Not only does it disturb a sense of distance and closeness; it indicates humanity's interconnectedness. We all desire touch yet are afraid of it too. We notice whoever touched us with a sense of "this is me and that is you." There is something distinctive about touch as the skin, our largest organ, responds to it. Energy from the person touching us is passed to us, as when Jesus noticed the touch by a woman in a crowd and the woman was healed, indicating the porosity of the body (Matthew 9:20–21).[38] When seeing, rather than looking or gazing, joins mindful touch, our coexistence is confirmed.

A final practice that informs traveling and facilitates meeting is *eating*. It is often when we eat with others that these unguarded moments of touch are likely to occur. The symbolic significance of this practice—of preparing food, eating, and sharing a table—cannot be underestimated. Travelers who are adventurous eaters and enjoy local cuisine know that there is an intricate relationship between food, table fellowship, and making new friends. Since traveling heightens one's sensory awareness, food and smells are experienced in powerful ways. A traveler's orientation to local food should not be confused with the tourist seeking the familiar golden arches of McDonald's, the patron on a cruise ship exploring an elaborate buffet, or cosmopolitan Westerners, often wealthy, enjoying ethnic cuisine. Cosmopolitanism seeks cross-cultural experiences too and becomes a badge of honor carried by many, creating industries such as food travel or culinary tourism and delivering phone apps to find reviewed local restaurants.[39] Due to numerous diasporas, emigration, and the relative ease of travel, eating ethnic foods is no longer the privilege of the well heeled or well traveled. Food enhances the level of an authentic experience. Lining up with local patrons and buying food from a street vendor or dining at a locally owned restaurant offers possibility. Travelers accept the invitation to table fellowship with the knowledge that the experience will add flavor to the food being eaten.

Orientalism's exotic impulse drives traveler and tourist alike to "authentic" and "unusual foods," such as eating "strange" animals or animal parts or tasting cured and fermented fruit and veggies. It is possible to approach food and eat in such a way that the distance between us and those we seek to meet increases and our stereotypes deepen. The slow food movement, with its emphasis on eating homemade food locally produced, taught us that local food ties one to local producers and their land, far removed from the corporate food industry with their industrial farms. When we eat local foods, we feel dis-placed as our outsider status is affirmed. It heightens the sense that we are guests and in a rather simple way invites cultural humility. The risk we accept eating food prepared on street corners adds to the experience and communicates physical and mental openness to new experiences. Sometimes local food is too strange, too alive, or too smelly to eat, and causes repulsion and disgust. Cultural humility encourages us to accept and decline invitations without feeling guilty or offending a host. If the saying "you are what you eat" is true and people eat what they are, travelers eat differently.

Actively seeking to meet the other, whether persons, communities, or nature, is a task travelers embrace. The traveler is a bridge (versus being a wall or a chasm) between locations of difference.[40]

The practices explored here assist travelers in changing not only their relationship with themselves but also their relationship to others and history. We "see beyond what divides us to what connects us, we're compelled to reach out beyond our walls of distrust, extend our hands to others and share information and resources. The survival of the human species depends on each one of us connecting to our vecinos (neighbors) whether they live across the street, across national borders, or across oceans."[41] Through traveling, we discover a new sense of belonging to humanity and to the created order.

The themes addressed in this chapter, from colonial powers to cultural humility to practices supporting meeting, are not foreign to the Judeo-Christian tradition. A few examples highlighting the themes we have thus far explored will suffice.

The Earthroamer and Mutual Recognition

We learn through God's self-revelation, both in the Hebrew Scriptures and around the person of Jesus, that seeing others is central to faithful living. Meetings of mutual recognition reflect the image of God alive in us, for God is an in-your-face God who created us into a facial existence. We long to see and be seen. God, speaking to Moses while organizing the religious life of the Israelites as they wandered in the desert, told Moses, Aaron, and the priests, "The Lord bless you and keep you; the Lord make his face shine on you and be gracious to you; the Lord turn his face toward you and give you peace" (Numbers 6:24–26). God's face seeks human faces and desires to be in relationship with us. It is no surprise that God consistently called on Israel to notice and take care of the stranger (Leviticus 19:33–34; Proverbs 5:10; Matthew 25:35). We too live a facial existence, a way of being that remains throughout the life cycle and regardless of our location. When we travel and experience moments of mutual recognition, the face of the stranger becomes the face of God and vice versa. Meeting someone and seeing the face of God is one way the Earthroamer surprises us on the paths of life.

However, seeing, as we have stated, easily becomes looking and gazing. In Luke 7, we read of Jesus visiting Simon, a Pharisee, who invited him for dinner. We imagine Simon's difficulty in meeting Jesus, as Jesus had already had a few tension-filled run-ins with the religious leaders around the interpretation of Scripture and the healing of people. The narrative indicates that Simon looked at Jesus but never saw him. While they were having dinner, a woman, a sex worker, learned that Jesus was visiting with Simon. She slipped into the home with a jar of perfume, possibly to give Jesus the perfume. Instead, she started crying, her tears wetting Jesus's feet. Touching his feet, she dried them with her hair. Simon, looking and gazing at the woman, possibly even disrobing her in his mind, was appalled, thinking to himself that if Jesus were a prophet, he should have known this woman's history. Reading Simon's mind, Jesus told Simon a parable of two persons with debt, one owing more money than the other. The debts of both were forgiven, with the one who owed the most exhibiting the most gratitude. Jesus then asked Simon, "Do you see this woman?" (Luke 7:44 CEB). The narrator does not answer this question, but the narrative indicates that the answer is no. Jesus then reminded Simon that he did not receive the customary foot washing and kiss when he entered Simon's home as a guest. Jesus forgave the woman her sins and told her to go in peace, acts that stunned Simon and all present. Jesus is asking all who hear about this event: "If you look at a person, whom do you see? Do you imagine restoration and wholeness for that person? Will you be fragile in your beliefs and stance in life and be affronted if goodness comes to the marginalized?" The Earthroamer sees a woman who longed for mutual recognition.

Another example that speaks to meeting comes when two disciples traveled between Jerusalem and the village of Emmaus, a journey of seven miles, a few days after Jesus was crucified and buried (Luke 24). As they walked, they recalled the horrific last days they experienced, including women discovering Jesus's tomb being empty. Suddenly Jesus joined them, but somehow they did not recognize him. Asking what they are talking about, Jesus feigned ignorance, and Cleopas, one of the men, continued to tell Jesus the events of the past days, half-blaming Jesus for not knowing about the events in Jerusalem. Cleopas shared how they were dismayed as the chief priests and leaders sentenced Jesus to death. He told their new companion that they all thought Jesus would reinstate Israel as a nation. Cleopas ended by revealing that an angel told women that Jesus was alive. Still, neither Cleopas nor his friend recognized Jesus. Jesus, as if surprised that he was not recognized, explained the Scriptures, telling about a Messiah who will die. Still, Cleopas and the other person remained clueless as to Jesus's identity. Entering Emmaus at nightfall, Cleopas invited Jesus to stay with them, which he did. Having dinner, Jesus broke bread and gave thanks, and suddenly Cleopas's and his friend's eyes opened. They recognized Jesus, who, in a strange irony, promptly disappeared—the meeting was but a brief *adventus* moment. With Jesus gone, Cleopas and his friend reflected on their walk with Jesus and remembered their hearts were burning, but they did not respond to their bodies' message. They got up and returned to Jerusalem to tell Jesus's followers what had happened.

This narrative reflects many themes mentioned in this chapter. We discover the Roman Empire and a people longing for a Jewish empire; we witness the subjugation of people; we see stereotypes tied to a person—Jesus—who operated at the margins. It is clear that when power and nationalist ideology join hands, injustice most often follows. We are struck that most of the narrative occurs while being on the road; we notice that food and fellowship lead to people meeting each other around new identities; we are told that moments of mutual recognition are fleeting. The narrative, which is an iconic "road story" in Scripture and plays a prominent role in the Christian calendar, leaves us with many questions: Why did Cleopas and his friend not recognize Jesus, someone they knew and saw just a few days ago? How do emotional states such as sadness and disappointment corrupt the ways we see? Is there so much discontinuity (alongside some continuity) in resurrection that one cannot be recognized? Why would breaking bread and giving thanks together open one's eyes? Why did Cleopas and his friend not listen to their bodies trying to tell them earlier that something is up? And why would Jesus promptly disappear the moment they recognized who he was?

Another example of meeting involves the apostle Paul, who was also called Saul of Tarsus. This meeting led to his conversion (Acts 9). Saul, a religious leader who persecuted those who believed in Jesus, was making his way from Jerusalem to Damascus, a journey of nearly two hundred miles, to turn synagogues against Jesus's followers. As he neared Damascus, a bright light appeared to him, with a voice asking, "Saul, Saul, why do you persecute me?" (Acts 9:4). Saul knew it was Jesus speaking to him. Those with Saul heard the voice but saw nothing. Saul, who had fallen to the ground, got up but found himself blind. At the same time, Jesus appeared to Ananias, a righteous man in Damascus, who was told that Saul was on his way. Saul and Ananias met at the house of Judas, who lived on Straight Street, and when Ananias touched Saul, scales fell from his eyes. Ananias reminded Saul that he would be filled by the Spirit. Having recovered from his Damascus Road experience, Saul became an apostle of Jesus, preaching boldly in Damascus and Jerusalem.

Paul's meeting led to a personal and spiritual transformation that baffled those who knew him. His actions changed from persecution to protection as he became an apostle of Jesus. As with the previous narratives, questions abound: Why did the men with Paul not see Jesus, though they heard his voice? Why did Paul go blind, but they did not? How should we understand Paul's confusion? What happened in that meeting that was strong enough to forever change Paul to new values and a new mission in life? Why did it happen on a road and not in the synagogues Paul frequented?

Seeking Restoration between Ambivalence and Awe

Seeing, listening, reading, touching, and eating establish relationship and form community. We have renewed awareness of gender discrimination, racial histories, financial inequities, educational differences, medical disparities, and other differences. We also experience appreciation, awe, and wonder. Emotionally, this feels like being on a roller coaster. On the one hand, we feel a deep connection with humanity. There is joy and open vulnerability, feeling content with things. On the other hand, we experience guilt, anger, and disappointment as historical tensions, inequities, and injustices remain.

Especially in nature and appreciating art, we experience *awe*, "a direct and initial experience or feeling when faced with something amazing, incomprehensible, or sublime."[42] It is "the perception of vastness that cannot be assimilated but can be accommodated, or as the experience of humility and wonder—adventure—toward living."[43] Awe is the deep awareness that one is small and insignificant and that one is experiencing something truly special or amazing. "Oh, wow . . ." slips from our mouths before we can catch ourselves. Awe awakens in us a deep sense of respect, admiration, and gratitude.[44] In a relationship, we feel deep connection. We also experience *wonder*, the "reflective experience motivated when one is unable to put things into a familiar conceptual framework—leading to open questions rather than conclusions."[45] Encountering beauty, the unexpected, the ineffable, the inexplicable, we think, "How do I fit into this complex world / expanding cosmos?" Wonder leaves us questioning, bewildered, disoriented, and curious.

Awe and wonder do not save us from witnessing the impact of climate change and knowing we've personally benefited from unjust systems. The ambivalence between emotional, relational, and spiritual questioning awakens a desire in the traveler to confess, to admit wrongdoing, and to commit to changed behavior. In the West, confession is a contentious practice tied to power structures of discipline, such as the church and the medical or health clinic, where it is mostly private, or confession is tied to the justice and penal systems, where it is public.[46] The desire to right wrongs is a human trait. Infants, before they have the verbal and cognitive capacities to process ambivalence, portray an awareness that things are not right, a confession of sorts prior to any sense of cultural morality. Around the fifth month, infants become aware that there is another person they just hurt, by either biting their nipple, pulling their hair, or poking a finger in their eye. A natural sense of guilt—which is really concern for the other person—sets in, leading to acts of reparation. Initially, the act of reparation might be a smile or going to bed without crying. For toddlers, it is coming up close to a caregiver, maybe touching their leg and obeying house rules. For adults, it is apologizing and admitting guilt; not pressing "send" on a hurtful text, email, or post; and living justly and walking humbly (see chapter 4). These acts portray *reparative intelligence*, the capacity to seek restoration for oneself and facilitating the restoration of persons, relationships, communities, and nature. It is being compassionate and caring.[47]

For travelers, acts of reparation, which reflect *conocimiento* and conscientization, come in varied ways.[48] An act of reparation includes significant self-reflection and assessing one's role in structures that oppress; it is admitting that one has been complicit in injustices and committing to change oppressive and unjust systems; it might be seeing and listening to and witnessing in parrhesiastic ways what others are experiencing; it is a public acknowledgment of privilege and power; it subverts power structures. Acts of reparation may be returning to the same community a following year to contribute to their communal living, or sponsoring a child in school through a nonprofit, or vowing to eat less meat and use less plastic to play a role in climate change. Acts of reparation demand a change in one's point of view and future actions to be more loving, more compassionate, and more just. Without such change and without action, the invitation for growth and mutual recognition within acts of reparation has been refused. With privilege comes responsibility, which requires acts that rewrite, undo, and correct all powers and relationships that undermine human flourishing. If acts of reparation remain abstract, they are mute.

An act of reparation is distinct from confession's need for forgiveness and receiving absolution, as the act implies changes in oneself first and then changes in one's relationship with others. To be forgiven, one needs someone to grant absolution, often someone with authority to do so. Forgiveness is a complex dynamic.[49] The colonizer, for example, cannot ask the colonized for forgiveness, as that would perpetuate the exploitation and power imbalance that demand attention to begin with. Whereas forgiveness often requires a long process of trust building and healing, an act of reparation is spontaneous and freely given, whether it is received or not. Acts of reparation deepen mutual recognition and shore up unintegrated ambivalences we experience when we travel. They bring new insights and awareness as they function beyond traditional power relations or religion. Ultimately, they bring a new *nepantla*-like identity to the traveler, who now bridges and engages the world with more authenticity and vulnerability, with an openness to being surprised.

Meeting a person or group and communing with nature, when colored by cultural humility and mutual recognition, are possible. Turning an encounter into an authentic relationship is an invitation that awaits every traveler. Certainly, overcoming distance and difference is difficult in an Orientalist, postcolonial world. Also, there is always a possibility that the boredom, loneliness, or frustration that prompted the urge to travel months ago will follow us into the journey. When we do not meet ourselves in surrender, we may not be able to meet others. Still, traveling seeks to redeem persons, humanity, and a creation straining under exploitative powers. Meetings of authenticity may be fleeting, but they may also be powerful enough to transform us and to awaken a new urge to travel. The irony that a life-changing meeting occurs in the presence of a stranger we are unlikely to meet again highlights traveling's paradoxical nature and the depths of our common humanity and of common grace. As we discover ourselves anew, meet persons, commune with nature, and experience awe and wonder, our attention and interest deepen—we admire, accommodate, and recognize a world in need of care and reparation.[50]

Chapter 4

CARING

Redeeming Privileged Irresponsibility and Building Solidarity

> Expert marching leaves no trails. . . . Expert reckoning needs no tallies.
>
> —Laozi, *Dao De Jing: The Book of the Way*

> But it is a chimera, this thing called tourism. It can offer fun, joy, rich existential experiences, ways to contribute constructively to conservation and to individual as well as social well-being. Yet there are also possibilities to wreak thoughtless harm on the destination, the environment, and those who inhabit them.
>
> —Tazim Jamal, *Justice and Ethics in Tourism*

"Take only pictures, leave only footprints" is a familiar aphorism not always followed. Travel remains a contested space as the chasm between abstract ideals—such as justice, equity, and sustainability—and concrete lives or ecological exploitation is not only easily exposed but also growing.[1] When we meet persons beyond an encounter, we discover more about their daily rhythms and lives, joys, and concerns. We realize our privilege and difference, which prompts a desire for a "confession," as we have explored. Descriptors such as "sustainable tourism" or "good tourism" or being a "responsible tourist" receive deeper meaning as we question our own travel habits. We are indeed in the time of the

Anthropocene, the time of significant human impact on the earth's geology and ecosystems. As we travel, we notice human footprints on persons made from topsoil, on the communities they form, and on the ground we tread.

Joshua Tree National Park in the southeast of California is a place of awe and wonder. Here, the Mojave and Colorado Deserts overlap, bringing a distinct ecosystem. Under relentless sun and with very little water, prickly plants and unique animals reflect adaptability, endurance, and fragility. The Joshua tree, the Mojave's iconic, waxy, and spiny yucca species, reaches for the sky, one inch per year. One is tempted to look up, though the presence of rattlesnakes, scorpions, tortoises, and iguanas demands looking down before stepping. Many visitors to the Park, however, do not care where they walk. Signs indicating cryptobiotic crusts ask all to stay on indicated paths. Yet footprints abound. Cryptobiotic crusts, also called desert glue, are cyanobacteria that grow hidden on the surface of the desert sand. The cyanobacteria grow filaments that are surrounded by a sticky gel-like sheath. As the filaments absorb moisture, they, with their sheaths, expand into a web of fibers, with sand and other small particles sticking to the sheaths. The fibers protect against soil erosion as they absorb the little moisture there is. Stepping on the cryptobiotic crusts leaves a footprint that remains, in some places, visible for up to several hundred years. Careless visitors even drive their recreational vehicles and trucks over the soil, leaving deep tracks. I stare at the footprints that crushed the cryptobiotic crust and wonder about the tracks I am leaving.

Whereas signs warn visitors to Joshua Tree National Park to stay on trails, there were none at Tenam Puente near Mexico's border with Guatemala. Tenam Puente is an underexplored Mayan archeological site. Our family decided to take a break from extensive language training in San Cristóbal de las Casas to explore the lesser-known heritage sites such as Tenam Puente and Chinkultic. These sites are barely excavated compared to the ruins at Zona Arqueológica Palenque. Our landlord recommended Miguel-Angel, a local guide, to take us to cenotes, used by the Maya as sacrificial sites, as well as various ruins. Dating from the sixth to the twelfth century, Tenam Puente is an active archeological site and noncommercialized. After visiting Palenque, the lack of a ticket office and vendor tables surprised us. Tenam Puente's pyramids and Mesoamerican ball game courts are impressive and force one to think of history and the rich culture that built these sites. Miguel-Angel educates us about the ball game—*ōllamaliztli* (or ulama)—where players used their forearms and hips to play with large balls made from rubber. The game reminds one of a mix of jai alai and soccer. We are joined by another group of three persons. With no barriers or signs, all of us are climbing up the pyramids for a better view of the site. As we walk the ball courts, we learn that some Maya sacrificed the winner of a ball game during religious festivals, as the gods deserve the best. We are climbing over and walking on sacred ground. There must be a better way to discover a heritage site.

The human footprint is large and growing, and depending on one's location, it is also very visible. Our family recognized this on a road trip through Alaska. We are in Whittier, arriving through the 2.5-mile single-lane Anton Anderson Memorial Tunnel that also serves the rail line. A sightseeing boat trip on Prince William Sound takes us close to a number of glaciers calving faster and forming bergy bits, small icebergs of glacial ice, all around our boat. A few days later, we are in Seward and walk up to the Exit Glacier in the Kenai Fjords National Park. The Exit Glacier is one of the Harding Icefield's forty-some glaciers. We are hiking through a cottonwood forest to a viewpoint where we see the glacier, which since 1815 has retreated 1.25 miles (2 kilometers). In 2013–14 alone, it retreated 187 feet (or 57 meters)! The rocky outwash plain it leaves behind is slowly being reclaimed by plants. On the way to Fairbanks on the Alaska Highway, the road is buckling due to the permafrost's thawing, turning the roadbed into a mushy mess. With calving and retreating glaciers and thawing permafrost, nature is trying to tell us something. Flying from Nashville to Fairbanks, our family added 6,328 pounds (or 791 pounds / 359 kg per person per flight) of carbon dioxide emissions to the atmosphere, not including the sport utility vehicle we used.[2] As we try to make sense of what we are experiencing, our family decides to stop using plastic bags when we buy groceries, to invest in solar panels, and to purchase an all-electric vehicle.

The human footprint can also be seen in sociopolitical and socioeconomic impacts on persons and communities, which remind us of our privilege. Having added even more carbon emissions by flying to South Africa to visit parents and grandparents, my partner and I take our preteen daughters on a township tour. Despite ambivalence, we feel the need to show our daughters how many South Africans live adjacent to cosmopolitan areas; how the legacy of apartheid remains. We discover a growing township tourism industry as townships are increasingly perceived as spaces of social and cultural value.[3] Our local guide, Themba, is a retired teacher who seems to know everyone. He takes us to a number of wood and tin shacks, to a *shebeen* (a bar selling homemade and other beer), a row of pit toilets, the one tap that serves many households, and a church that has a day care service as well as an after-school program. The latter is needed, as many mothers work in white households in the suburbs. At the day care center, we meet a young woman from Sweden volunteering there during a gap year. She lives in the community. Themba apologizes for the mountains of litter, saying that the city's sanitation department poorly serves this neighborhood. We feel a vibrant energy, witness a strong sense of community, and experience creative entrepreneurship with *spaza* shops (informal grocery stores) everywhere. These forces are challenged by long-standing social and economic injustices, unemployment, and high crime rates. We are deeply touched by what we experience. Themba reminds us that his name means "hope" and that he is indeed hopeful about this community's future. The days following the visit are filled with conversation and discernment as we continue our road trip. We decide to sponsor a girl's education and would later receive reports about how Restituta is excelling in her studies.

Justice as Distribution

There is no traveling without predicaments, ironies, and ambiguities. If justice expects that "all social primary goods—liberty and opportunity, income and wealth, and the bases of self-respect—are to be distributed equally unless an unequal distribution of any or all of these goods is to the advantage of the least favored," as Harvard political theorist John Rawls writes as he explores a theory of justice, two definite predicaments a traveler faces are inequality and injustice.[4] Knowing this, travelers choose their lodging carefully, attentive to matters of ownership and the wages service personnel receive. When taking a shower or bath, we remain mindful that visitors, in general, use more water than locals, especially if the place we visit faces drought. Some travel to places where human rights are denied, especially to women and girls; to persons of diverse sexual orientations; to the elderly, the homeless, and the differently abled; and to persons of minoritized religious affiliations. Others refuse to visit a country in protest against the governance structures and its poor human rights records.

Many travel to provide medical and social services governments deny their own people. In doing so, they help many locals, yet their helping also keeps the government from providing care to its citizens. Travelers remain aware of the food they eat and minimize food waste. They note when healthy and nutritious foods are often reserved for tourists and travelers, when the local community eats foods less nourishing because they cannot afford the food reserved for visitors. Hikers stay on marked paths and carry their waste out. Others travel for months on end, immersing themselves in local communities and economies for periods of time before moving on. Many travelers watch their carbon footprint, live simply, and consume differently to prepare for their next adventure.[5] All travelers, however, know that calling for an end of travel or tourism is not a reasonable option, as 10 percent of the world's gross domestic product and one in every ten jobs are linked to the travel and tourism industries.[6] It is no surprise that voluntourism—combining travel with addressing a cause such as alleviating poverty or combating climate change—is a growing industry.

Traveling has many positive impacts, such as "the creation of jobs, economic diversification, revenues and foreign exchange, new services (e.g., restaurants), and facilities (e.g., for events, recreation), and social benefits like increase in community pride and tolerance," while it also has "negative impacts like low-paying jobs (rather than a living wage) and limited career advances, the quickening of gentrification, seasonality challenges (where peak visitation periods place immense stress on environmental and human resources), as well as the commodification of culture."[7] The traveler can overcome the paradoxes of the tourist industry by *caring*. We face concrete situations that call for a response. We seek to be ethical travelers who are spiritual, religious, and moral agents, limited in what we are able to do in foreign worlds but not without agency either. It is this caring response that we seek to explore in this chapter. If we do not have a caring approach to travel, it is possible that we travel in such a way that we maintain or even enhance the divisions of injustice. The complexity of this world denies a universal response a traveler might seek to adopt. Notions of being a global citizen seeking global justice are too abstract and fail us. Still, with critical skills, mindfulness, and a caring approach to life, travelers can contribute in positive ways, entering the social, cultural, spiritual, financial, and natural environments that communities and destinations offer.

Travelers have to overcome their "environmental bubble," which keeps them focused on their home experiences and also disconnected from local cultures and lifestyles as they constantly compare what they know with what they are experiencing.[8] Still, we can and do question the dynamics of power, of privilege, of supremacy, and of the hierarchical structures we witness.[9] It is possible to travel with sustainability in mind and to travel justly, to "[meet] the needs of present tourists and host regions while protecting and enhancing opportunity for the future."[10] As we follow the Earthroamer, we seek to live well and to participate in building a quality life for all; we protect and sustain creation. Our care runs deep not only because we have a natural desire to restore what is broken but also because it includes loving oneself, loving neighbors, and loving the Earthroamer and all of creation. Caring in a world of growing inequality demands intention, even courage.

We will explore caring by arguing that empathy as an emotional skill is insufficient; that caring has phases or gradients of deepening intensity that invite responsibility and solidarity; and, for those following the Earthroamer, that caring is a spiritual mandate that deepens relationships we form with individuals, with communities, with nature, and also with God.

When Empathy Fails

Empathy is often seen as the ability to imagine oneself in someone else's shoes—a humanizing action. In theory, empathy sounds like a mature, moral response. In a colonized and colonizing world, a world that is driven by neoliberal capitalist and transnational economic policies, however, empathy most often fails because it leads to very little or no change for the poor, the oppressed, and the marginalized. Neoliberalism describes the political and economic policies that began in the mid-1980s as a response to the civil and women's rights gains of the 1960s and 1970s.[11] As a reaction to this greater inclusion, neoliberalism is associated with free market capitalism, deregulation of industries, globalization, trade agreements, tax cuts (especially for the wealthy and corporations), and shrinking government sectors while private sectors of the economy are supported and built. The result of this ideology is that companies serve their stakeholders only, and little of their profits flows back into communities. In a neoliberal society, public institutions, such as libraries, schools, and hospitals, receive little or no financial support from federal, state, and local governments. Those of us who are salaried see our income not keeping pace with inflation or the cost of living. Slowly we are migrating to the periphery of the economic landscape. The economic marginalization of neoliberalism has real-life and social implications and is balanced by the privatization and commodification of personal and public assets. Large corporations make money off us under the guise that they have our best interests in mind. Of concern for us is that especially international tourism is underpinned by an implicit and often explicit belief in the sanctity of the free market, open borders, and the deregulation of markets. The tourism industry and neoliberalism are intimate partners.

No longer a term assigned to left-leaning academics, neoliberalism is directly impacting the everyday lives of the majority of people on earth. Due to deregulation policies, the earth is being exploited, more than ever, as seen in deforestation, the production of beef and palm oil, and turning sand into a commodity. As travelers, we are interested in the real-life implications of neoliberalism alongside its ideological foundations. Add to the neoliberal landscape the impact of the gig economy—the economy built around independent contractors using internet platforms with few worker benefits or protections—and we have powerful movements against inclusion and equality. Much is reduced to the private sphere under the banner of "freedom." In this world, empathy, which opens one to the experience of others, fails, as it rarely leads to meaningful action and systemic change. We are reminded of civil rights leader Dr. Martin Luther King Jr.'s call to embrace an "ethic of love": "[One] seeks to defeat the unjust system, rather than individuals who are in that system. And that one goes on believing that somehow this is the important thing, to get rid of the evil system and not the individual who happens to be misguided, who happens to be misled, who was taught wrong. The thing to do is to get rid of the system and thereby create a moral balance in society."[12] When we travel, we do not leave our ethics at home but seek to be part in creating a moral balance in all societies. But how should we think about empathy?

We can learn from Carolyn Pedwell, a British sociologist who explores the international politics of emotion and embodiment. "Indeed, within the neoliberal context in which international development operates," she writes, "empathy can become an emotional skill or capacity with both institutional and market value."[13] Being empathetic, a skill gained through immersion experiences and training, does not inevitably lead to action that betters the lives of the poor and the marginalized. Rather, as seen in international development programs, empathy may lead persons to feel they know "the truth" about people's lives and struggles. The voices of the poor and marginalized are muffled yet again. Empathy, a prized Western value, holds the promise to overcome chasms of difference but becomes a liability for the poor.

Pedwell names how empathy is seen as an affective skill that is cultivated especially in middle-class white Westerners: the privileged. Men seek to gain this skill and are compensated for it. Women, in turn, find that their empathetic skills are rarely recognized or compensated. Empathy is easily universalized—we imagine "the truth" about the poor and what they are experiencing while ignoring historical, economic, and contextual issues. This kind of empathy has been called "cosmopolitan empathy . . . which refers to people's ability to adopt the perspective of geographically distant others and to feel compassion for their plight."[14] It speaks of a capacity to feel someone's or a community's plight but to do so in self-indulgent ways, to exploit people further, or to remain indifferent. Cosmopolitan empathy benefits the person who is empathetic more than the persons or communities who are being empathized with.

Whereas care is always caring *with*, empathy as an affective skill most often leads to the poor non-Westerner becoming the *object* of empathy and care. The empathetic Westerner now *chooses* to be empathetic or not, assuming a power position yet again. This choice is affectively made. Research indicates that children and persons with a visible disability, for example, receive more empathy than women, the elderly, or those with invisible challenges. Neoliberalism renders empathy a skill to maximize economic competitiveness and growth, processes that are selective in terms of who is engaged and that leave the poor behind. Critics of empathy remind us that the emotion may have been reduced to affective knowledge and understanding of the other—"I know (or imagine) how you feel"—as the promise of more ethical action and increased social justice never comes to fruition.

The challenge we face with empathy as imagining how another feels or what they are experiencing is that we have to be more intentional to escape the geopolitical histories and relations and the neoliberal capitalist economies that shape our world. Knowing how a packaged tour backed by a corporation trickles money to the local economy, we choose to use local guides instead.[15] We join movements that combat the exploitation of workers or that seek climate justice. We seek out justice tourism that addresses inequalities and injustices. Likewise, we choose not to travel to certain places due to their ecological fragility or in solidarity with locals who called for a moratorium on travel. Whether these choices are enough if we never have to give up our power and privilege remains a question.

Rather than simply being empathetic and entering into sentimental attachments, *we care*, an action that relies on empathy as it engages concrete concerns and fosters relationships of solidarity. Care moves beyond engaging objects of empathy from a distance; instead, it demands proximity and intimacy.[16] Caring recognizes one's own complicity within transnational hierarchies of power and compels working with others toward systemic changes. Caring is open to being personally affected by the experience of the other and draws on one's reparative intelligence, which facilitates the restoration of lives and landscapes as it resists maintaining unjust social and economic practices and policies by governments and corporations.

Caring: A Way of Being, Seeing, and Doing

Travelers who follow the Earthroamer use their empathy to care. Though caring is such a natural practice, it is, much like traveling, vulnerable to the sociopolitical and socioeconomic landscapes we call home. "Care," political theorist Joan Tronto states, "implies reaching out to something other than the self; it is neither self-referring nor self-absorbing. Second, care implicitly suggests that it will lead to some type of action."[17] To care is a "way of thinking" and "a way of seeing a society" that leads to "doing."[18] When a person states that she cares about world hunger but does not do anything about it, we think the person did not mean what she said and does not really care. Care, from the Latin *cura*, has connotations of being burdened; being concerned and having responsibility; it is weighty and even an inconvenience. Caring is not simply showing interest or participating in a thought experiment; it is being troubled into action. Tronto's definition of caring complements surrendering and meeting as core aspects of one's travel experience: "Caring [is] *a species activity that includes everything that we do to maintain, continue, and repair our 'world' so that we can live in it as well as possible.* That world includes our bodies, our selves, and our environment, all of which we seek to interweave in a complex, life-sustaining web."[19]

We care to overcome our cosmopolitan empathy. Tronto envisions acts of care with persons and things, including the environment. These acts are neither individualistic nor dyadic but rather embedded in and springing forth from a web of relationships. "Caring with" always has a place, a people, and a location in mind and, as such, is defined culturally. It is also an ongoing action, as care is needed throughout the life cycle for humans and at all times for nature. Our caring is limited only by our capacity to build relationships and our imaginations' skill at addressing challenges. Tronto warns that we should not confuse caring with other actions such as the pursuit of pleasure and fulfilling desire, creative activity, production and marketing of new products, and acts of destruction. Some actions, such as playing or protecting or participating in rituals seeking to minimize violence, are both caring and have other ends in mind. Traveling is such an action, as its multifaceted nature has more than one end in mind, one end being a caring and just society. "Caring seems to involve taking the concerns and needs of the other as the basis for action," Tronto writes.[20]

When we travel, we seek to make a difference, enter into relationships with persons and communities to see them do as well as possible, and experience a renewed and sustaining relationship with nature. We do so by caring about, taking care of, providing care (caregiving), receiving care, and caring with others.[21] *Caring about* involves becoming aware of and paying attention to the need for caring. Westerners often see caring about as an individual matter—caring about someone—when we have learned that caring about social and political systems will have a greater impact on a larger number of people. Visiting a local organization that is addressing need is possible for most travelers. Few of us have deep pockets to write large checks, but we can offer others our persons and our time; we can promote their work. To care about takes time and intentionality and is not always driven by money or extensive resources.

Taking care of is a next step in the caring process. This step assumes that one takes responsibility to meet a need that has been identified. It accepts the fact that one is able to make a difference, even as a traveler. Here too we have to resist the temptation to focus on an individual. One might focus on the needs of women and girls, for example, as that will impact boys and men too. Since taking *care of* is also *caring with* others, as we will explore soon, caring crosses divides between persons and in communities, and through care, the possibility of human flourishing is introduced. We overcome hierarchies of power by listening to the voices of those at the margins, a practice we start when we listen beyond an encounter. Tronto warns that if we have the mindset that giving money to an organization is of no use, as the money will fall into corrupt hands, we also admit that the need cannot be met and thus cooperate in maintaining the status quo. Partnering with organizations is the easiest way a traveler turns *caring about* into *taking care of*, which leads to caregiving as a next phase of caring.

Caregiving involves the direct meeting of needs for care. It is joining a team that provides medical services and fulfills other needs, it is interrupting one's travels for a few days to paint a school or to build a portion of a clinic alongside local workers and stakeholders, it is volunteering at a service organization and handing out food, and it is helping a nature conservation agency clean streambeds or protect turtle eggs. Giving money to an individual, Tronto states, is a last resort, as that person will have to turn the money into something else to fulfill their need. It is akin to the traditional breadwinner "taking care" of his house and earning a living, expecting his wife or partner to turn those earnings into a cooked meal, a clean house, a washed load of laundry, an organized social life, and more. Tronto wonders whether this man can take care of himself. For many, this traditional way of dividing labor in the household remains. Whereas giving money is a form of *taking care of*, it is not *caregiving*. Since joining a service organization may take some planning, travelers design their trips accordingly and may reach out to a service organization in advance. We research how a personal interest is present and being addressed in the community we will enter. Joining a service organization places one in *a process of care*, which is almost always better than a one-off, individual act. It speaks to the work of all and not the work by an expert. One can begin to build a bridge to communities and societies before leaving by seeking referrals from churches, other faith organizations, and charities, and of course by searching online.

Care receiving is the next step of the process of care. This step involves witnessing the response of the person, group, or community who received the caregiving, which can range from expressing gratitude to indifference. Care receiving implies putting oneself in a position where the results of care are witnessed. On a medical missions trip to Honduras, I was a chaplain and a gopher. While surgeons, physicians, dentists, and nurses provided hands-on medical care, I handed out medicine, made sure those waiting in line had water, and prayed with persons. One moment that remains with me is experiencing the gratitude, through laughter and tears, of a man who was blinded by cataracts for more than thirty years. Two days before, he received eye surgery and, on this day, he was visiting with the surgeon to remove the eye patches. His ten-year-old granddaughter, charged by the family to take care of her *abuelo* and thus unable to attend school, was standing by. As the patches came off and after drops were placed in his eyes, our *abuelo* smiled, then laughed, and then cried. It was the first time he saw his granddaughter, whose life too suddenly changed! Through acts of care, miracles yet happen and the blind do see again.[22]

A fifth and final phase of caring is *solidarity as the act of caring with*.[23] This aspect of caring celebrates its deeply relational qualities and the fact that it is built on plurality, trust, and respect as well as communication. The local contexts always determine what *caring with* looks like as it anticipates making a collective response to perceived needs and a common resistance to neoliberal commodification. Caring travelers go beyond just experiencing new relationships, communities, and places. Rather, they partake in the restoration of all of creation. "Care," Tronto concludes, "is a power of the weak," for caring is essential to supporting life.[24] The relationships formed through caring are often sustained, as caring demands facial recognition, which we discussed in the previous chapter. The otherness of a person is removed through our caring; we are united in our joint humanity. Being united speaks to solidarity and determines one's relationships with all of humanity, cultures, geographies, and ecologies. "Solidarity," a word that only entered the English language in 1841, is "the fact or quality, on the part of communities, etc., of being perfectly united or at one in some respect, esp. in interests, sympathies, or aspirations."[25] Solidarity calls a traveler and a host together to make a difference.

Through *caring about*'s attentiveness, *taking caring of*'s responsibility, *caregiving*'s competence, *care receiving*'s responsiveness, *caring with*'s trust and action, and *solidarity*'s empowerment and connection, we care well, even as travelers. It takes thought and action, yes, but caring enriches the experience of any journey. Tronto warns that caring is never easy, is always multifaceted, and involves conflict. The last stems from different needs that invariably arise. What need will be addressed, by whom, and in what manner? Caring when we have met someone adds a relational component of mutuality that minimizes the negative fallout of conflict. The conflict inevitably also implies tackling the neoliberal policies that keep services away from people under the guise of freedom and personal responsibility. Neoliberalism's lack of attachment is clear as society's safety nets for the poor, differently abled, or marginalized are removed or portrayed in negative terms. Inevitably, caring is wholly political. When we see caring as a way of being, seeing, and doing, our traveling practices change accordingly.

Engaging all five phases of care might not be possible on every trip, but some aspects of a care ethic inform one's daily practice, even on the road. We bring our care to the personal, the social, the cultural, the economic, the political, and the environmental spheres of life, for escaping these playgrounds of life is impossible. Caring, as we have explored it here, is concrete and contextual; there are no universal truths. It calls us beyond individualistic thinking and empowers us to make a difference as we travel.

To Care about Personal Safety

One element inherent to caring is self-care, which includes protecting one's personal well-being and also safety. Travelers know that traveling carries risk. It's a fact that flying is safer than driving and driving in some countries is much more dangerous than driving elsewhere. A potholed road shared by hitchhikers, ox wagons, large trucks, buses, cars, and even wild animals cannot be compared to interstate highways or Germany's autobahn (large sections of which were built by Nazi prisoners in World War II concentration camps). Infected water or bad food gives one a bout of traveler's diarrhea, which is always a humbling and vulnerable experience, and dangerous if the dehydration cannot be stopped. Travelers face insect, reptile, and animal bites; natural disasters such as fires, earthquakes, and tsunamis; even the risk of kidnapping and attacks from religious or political radicals. For many, these moments describe their home experience. We get inoculations or take prophylactics for illnesses ranging from malaria and yellow fever to hepatitis A and COVID-19. Travelers who are sexually active, research indicates, often compromise safe sex practices.[26] We also practice digital safety, making sure we use a secure Wi-Fi signal or a virtual private network when we travel, else risk a virus of another kind. Though popular tourist destinations often have a higher level of police and security presence as local and state governments protect their tourist industry, we critically engage this securitization and ask: *Protection from whom? Security for whom?*[27]

The West's colonial roots instill fear of the stranger, of the poor and the pitied, of the person of color, or of whoever is deemed the other. In turn, persons of color are vulnerable to racist experiences when traveling. Traveling alone as a woman increases one's risk for sexual harassment or worse, especially when visiting countries with a poor track record of protecting girls and women. Persons with different abilities know that many countries are poor at implementing accessible design. As we seek to meet persons, we remain mindful of our privileges and prejudices and work hard to overcome any racist and xenophobic roots. Though the Central Intelligence Agency (CIA) serves the United States' mission to remain a dominant player on the world stage and has a checkered history with their political involvement in other countries' affairs and abusive interrogation techniques, the CIA's *World Factbook* and the *Factbook*'s *GINI Index* measuring inequality give a quick introduction to a country. These and other forms of information are helpful in assessing risk and providing information on geopolitical matters—as well as embassy and local emergency numbers.[28]

At best, traveling is a calculated risk. One would think that keeping the risks in mind would stifle the traveling spirit, and it does that for some. However, any traveler will be quick to say that the reward of travel far outweighs the risk, as no place is risk-free. Finding a place too dangerous to visit challenges the practice of caring with its restorative emphasis, as we have explored. Furthermore, individuals and families with children, sometimes our own family members, call those places we question home. Easily dismissing a place often reveals our privileged location or status. Instead, we should care through actions of solidarity.

Overcoming "Privileged Irresponsibility"

In the practice of traveling, no one gets a pass not to care; placing care at the center of the practice of traveling brings possibility. One outcome is the move from traveling as *economic production* to traveling as *the production of care*.[29] Since traveling is always a privilege and tied to neoliberal capitalism, how we travel helps us overcome the danger of "privileged irresponsibility." Tronto writes, "Those who are relatively privileged are granted by that privilege the opportunity simply to ignore certain forms of hardships that they do not face; I suggest we call this form of privilege 'privileged irresponsibility.' . . . Dividing up responsibility privileges those who are excused by not needing to provide care; thus the privileged avoid responding directly to the actual processes of care and meeting of basic need."[30]

Tronto is addressing how some in society, the wealthy and men most often, receive a "pass" or a cultural validation or excuse not to care.[31] Traditional gender roles, for example, push care onto girls and women or those who are less educated. Though society says caregiving is important, those providing care are devalued and poorly compensated. Through financial and other means, we insulate ourselves from the injustices and inequalities of the world. Moreover, privilege anticipates being served and using others to meet personal needs. When we hire others to do services for us, we may see that as providing care to them because we compensate them for their service. Caring is much deeper than a financial relationship, however, as we have learned. Privileged irresponsibility is part of white privilege and, like racism, blinds some to the needs of others. And when a need is recognized, the responsibility to meet that need is placed on the person who has the need. Neoliberalism believes that every person can pull themselves up by their bootstraps, and the poor and marginalized should do that so as not to be a burden to society. Privileged travelers who claim no responsibility toward the care of others locate themselves on the extreme end of the caregiving horizon where detached observations are made. They reveal their privileged irresponsibility. The community paying their workers the going rate but not a living wage, as we learned in the previous chapter, was unaware of their privileged irresponsibility. My attempts to escape this unjust position may have been insufficient.

A term such as *responsibility* is often used when we explore morality. Neoliberalism politicized the term and places responsibility on those in need or on those who lack financial security. When we bring notions of responsibility and irresponsibility to the practice of travel, we acknowledge that there is something deeply personal, moral (or ethical), and spiritual (and religious) about leaving home. Our exploration of care, however, shows that we are response-able—we care, even in our travels. Letting go of our entitlements is possible. The care we portray while traveling denies us any abstract understanding of needs such as poverty, food scarcity, lack of education, or medical services. We meet persons who witness those needs to us. When we embrace caring solidarity as a way of being that calls us into action, traveling becomes an opportunity to overcome the risk of privileged irresponsibility and its attitudes of superiority and exploitation.

The danger of privileged irresponsibility is that it functions in unconscious and invisible ways. Certainly, it is possible to visit communities and destinations without caring. When we look and gaze and observe the vast need around us or the impact humanity has on nature and walk away in detached fashion, something is off. Since meeting persons raises our awareness of need, remaining in a position of silent privilege, misinterpretation, or even ignorance is not possible. Meeting is built around relationship, vulnerability, and equality, values that break through privileged irresponsibility's walls. Those in the relationship have equal voice, which is power. When we choose not to respond to the need we witness, we have to harden ourselves, which impacts us in other negative ways. We will return home with that detached self, our loneliness fueling deeper levels of individualism and alienation.

Just Traveling sees all travelers as politicians, but not in the ways that "politics" is viewed in contemporary society. Plato, in his *The Republic*, has Socrates define the ideal city's constitution, as that would shape the city itself. The original meaning of "politics" is thus "the shape of the city," which includes the city's laws and rules for governance but also its roads and architecture, its poetry and music, its infrastructure, and the arts.[32] All these factors shape a city. As Jesus did, we shape our travels and choose the communities and places we visit around values of compassion and care, around the grace of loving self and neighbor. When politics today never wanders far from *who gets what*, caring compels us to ask *who needs what* and *who does what?* No one gets a pass not to care, travelers included. Caring as we explore it here is about intimate and daily concrete routines such as going to work or driving across town, but it also anticipates acts of hands-on care done with compassion and without judgment away from home. As such, caring is deeply spiritual and liminal. In the overlap between self and another, in circling a cause or those structural problems we need to address to ensure a flourishing life for all, we meet self, other, and God anew. Seeing a blind man gain sight is not easily imagined and changes one's view on life in significant ways. Reading how Jesus gave sight to the blind is now reframed too. When many in society take a pass on caring, travelers who care bring possibility to effect change wherever we go.

Deep Solidarity as a Spiritual Practice

Whereas few will argue against caring as a spiritual practice, *solidarity*, with its political and revolutionary French roots, has only now been discovered as a spiritual practice worthy of theological exploration. Theologian Joerg Rieger in his *Faith on the Road* argues that "Christianity is a matter of the road."[33] He reminds us that Jesus's ministry most often took place on or alongside roads and rarely in homes or closed spaces. "Traveling can help us broaden our horizons in important ways that are transformative not only for the world but for the church itself," he writes.[34] Through our traveling, we connect with a core aspect of the Judeo-Christian tradition: that it is a tradition of people on the move; that neither the tradition, nor its Scripture, nor its followers remain static. We also discover God anew as "God looks different from the dusty roads of Galilee than from the safety of the temple in Jerusalem. Today, God looks different from the street level of our cities than from the corner offices of economic, political, and religious privilege."[35] As we discover persons on the move and people meeting Jesus on the road, there is a silent warning for those not on the move, those who rest comfortably for too long, and those who do not anticipate meeting Jesus where they travel. Rieger's observations regarding movement in tradition and Scripture inform our thinking as we explore travel as a choice made in freedom to leave one's home with every intention to return. He reminds us that travel is also "tied to tension and displacement"; that there are tourists, pilgrims, and vagabonds (those moving in freedom for extensive periods of time), yes, but also migrants, refugees, exiles, and the dispossessed on roads seeking freedom, people for whom leaving was not a choice and returning home may be impossible.[36]

A seasoned traveler, Rieger describes the Earthroamer God in Jesus, who spent so much time on the road that the church created a *christologia viatorum*, a Christology of travelers and pilgrims, and a *christologia viae*, a Christology of the road.[37] Rieger too is concerned about meeting the other in a neoliberal capitalist world of economic inequality and a God too closely identified with empires. As Rieger identifies faith communities that are stuck in place, he faults their mission and immersion trips, which never broaden any horizons: "Those with money can insulate themselves from the hardships of everyday life not only at home but also when traveling abroad."[38] As we argued, having empathy toward the poor and disenfranchised and linking social issues with neoliberal capitalism are not enough. Travelers care in concrete acts!

Since mindful travel does not allow us to remain neutral, *faith on the road* encourages us to embrace "deep solidarity," an awareness that the financial systems of this world work for a few wealthy only.[39] It is a call to recognize the differences among us and to work together for a common cause. Solidarity overcomes the divide-and-conquer tactics we have identified in maintaining and expanding empires, which are rampant in sexism, racism, religious homogeneity, and other forms of discrimination. Through solidarity, we engage the world beyond thought and advocacy and turn travel into an act of justice, a spiritual and political act.[40] As we travel, we seek to change concrete inequalities and hear the voices of others—resisting the power inherent in "giving voice to . . ." We recognize the "soft power" the West and neoliberal capitalism use to assimilate people into a white way of seeing the world without these powers giving up any cultural, political, or economic superiority. We identify anew with the vast majority of persons who are workers. We find fault with market economies that subjugate people and the natural world to global exploitation, arguing for fewer rules to regulate matters while commercializing persons and nature.

Deep solidarity on the roads of life is solidarity *with* others, solidarity that embraces and celebrates vast differences between peoples without trying to erase those differences. Our world is desperately in need of this deep sense of belonging to a diverse humanity. As we return to a community a following year and our relationships strengthen, we are part of the bend toward justice we want to see in this world. Both caring and deep solidarity ask: *Do persons of faith travel differently? And if so, what difference does their travel make?* The answers to these questions vary but will not wander far from values such as peace, justice, equality, and care. We embrace travel's invitation to move from affective dissonance in the face of injustices and inequalities to affective solidarity and getting involved in concrete ways through surrendering, meeting, and caring. Scripture, after all, calls us toward this vision.

Scripture's Call to Do the Right Thing, Wherever One Goes

From the Good Samaritan we learned that caring is a road experience of surrendering. An ethic of care is often built around Scripture's call to *love one's neighbor* as oneself (Leviticus 19:18; Matthew 19:19, 22:39) or to mirror God's *compassion* (Exodus 34:6; Joel 2:13; 2 Corinthians 1:3). Either of these interrelated foci for grounding an ethic of care is sufficient. Looking at a few texts addressing our walk with God sheds some additional light on caring as a core aspect of the traveling experience and its spirituality.

Micah, the eighth-century BCE prophet, was the first to predict the downfall of Jerusalem. He also promised restoration for Jerusalem and prophesied an era of universal peace. Seeing how leaders spend money to beautify the city while its citizens remain poor, Micah tells the people of Jerusalem and neighboring Bethlehem, "[God] has shown you, O mortal, what is good. And what does the Lord require of you? To act justly and to love mercy and to walk humbly with your God" (Micah 6:8 CEB). Every traveler has this call tattooed in invisible ink on their forearm. The text imagines God taking Israel to court for not being faithful to God's covenant laws. In the preceding verses, the Earthroamer, like a wise prosecutor, asks the people of God how they were burdened as they were freed from Egypt. God recalls the Exodus narrative and other moments of salvation. Unable to challenge God's case against them, Israel seeks to reaffirm their covenant with God. They wonder whether offerings of precious lambs, olive oil, and thousands of rams would suffice. As the offerings escalate in value, we sense Israel's desperation to atone when they contemplate sacrificing a firstborn child. The offerings of atonement of the people of Jerusalem fail, possibly to their relief—for who wants to sacrifice their firstborn? It is then that they hear that God wants them to act justly, to love mercy, and to walk humbly.

To act justly is to abide by God's laws, those guidelines that ensure concrete personal and communal well-being and a relationship with God. This justice anticipates restoration for persons in poverty for whom a flourishing life is all but a dream, for persons whose rights and opportunities have been removed by mismanagement or an oppressor, for persons who are marginalized and without power. To love mercy is to remember the grace the Earthroamer flows to humanity and all of creation. It is to step into that same stream that gives life without judgment as it flows to barren lives and places. Mercy calls one to go beyond the call of duty, to go the extra mile. To walk humbly, as the Hebrew term *hasnēa* (meaning "humility" or "being humble") suggests, is to walk mindfully, with care and caution, without arrogance. In humility, we resist placing ourselves above others. This humble walk also carries connotations of being ready, to anticipate a person in need of justice and mercy and to be responsive to that person's need.

Micah 6 speaks to portraying certain practices and embodying certain attitudes. We explored those in terms of caring and solidarity—ways to not only walk *before* God but also walk *with* the Earthroamer. Our "vertical" relationship with God and our spiritual vitality are inextricably linked to our "horizontal" relationships with all of humanity and creation. One need not travel vast distances or wait for special moments to act justly, to love mercy, and to walk humbly. With traveling and caring built on relationships, we encounter need in every person we meet. Our very neighborhoods need God's restorative work and our participation in that work. Whenever we leave home to explore new communities and destinations, whether the other side of town or the other side of the globe, this call from God to act justly, to love mercy, and to walk humbly rings in our ears.

Jesus, the Earthroamer incarnate, informs our discussion on traveling-as-caring too. In Matthew 9, we discover Jesus as he heals a paralytic and two blind men, calls Matthew as an apostle, educates John the Baptist's disciples on fasting, raises a young girl from the dead, and casts out a demon in a mute man. Then we read, "Jesus traveled among all the cities and villages, teaching in their synagogues, announcing the good news of the kingdom, and healing every disease and every sickness. Now when Jesus saw the crowds, he had compassion for them because they were troubled and helpless, like sheep without a shepherd. Then he said to his disciples, 'The size of the harvest is bigger than you can imagine, but there are few workers. Therefore, plead with the Lord of the harvest to send out workers for his harvest'" (Matthew 9:35–38 CEB).

Matthew 9, which begins with Jesus's teaching, preaching, and caring, ends with his concern for the world. It is early yet in Jesus's ministry, but he is frustrated and concerned. Though the people reacted to his miracles, they seemed not to grasp his message that the reign of God has come near through his person, message, and works. Witnessing the crowds, Jesus is overcome by compassion. He sees a people—Israel—in physical need and lost in spiritual ways. The Greek suggests that Jesus's innards were torn by what he saw. This is the burden a God of compassion carries. Jesus then draws on the Hebrew tradition of the harvest, which became symbolic of the end times when God would restore order to this world and the angels would reap the harvest. Our text is important as Jesus envisions his followers going into the world and doing exactly what he was doing. As the disciples changed people's lives, they became Jesus as the Christ incarnate to others. Travelers who care in solidarity incarnate Jesus too.

The text is not without problems, as missionary zeal joined empires in bringing light to persons in darkness, judging local traditions and religions harshly. So too those who are labeled sheep are typically seen as dumb and in need of a leader. Sheep are passive and voiceless, far removed from being equals in relationships of mutuality. These historic misinterpretations of Jesus's words have caused much harm and death in the world and are a stain Christianity constantly has to address. Jesus, however, does not make his name and a relationship with him a litmus test of anything. As Matthew 10 states, the mission of the disciples is not to convert people but to heal and to bring transformation. There is a sense of mutuality and even solidarity. We remember the parable of the goats and the sheep that Jesus told near the end of his life (Matthew 25). Here too salvation is not tied to a relationship with Jesus but to giving food to the hungry, water to the thirsty, hospitality to the stranger, clothes to the naked, and care to the sick. Jesus notices the poor, the needy, the oppressed, not to further exploit and marginalize them or to offer them a spiritual salvation only, but to welcome the marginalized back into humanity's fold by tending to concrete needs. These acts cannot be confused with "saving" people from hell or "confronting" people with Jesus. Jesus's followers are not asked to speak about things that do not impact concrete need but instead to do good. Matthew's crowds are people impacted by oppressive systemic structures that do not reflect the reign of God. Our wander with the Earthroamer mirrors Jesus, who traveled among cities and villages to recognize those often overlooked. We need not be concerned by what we will say but be creative in the ways we will join others as we do something to alleviate need. We follow the Earthroamer not out of piety but to embrace the call to be someone and do something countercultural.

Another text aimed at us travelers is found in the Epistle of James (4:13–17 CEB), where we read what has become known as the *conditio* (or *reservatio*) *Jacobi*, or Jacob's reserve, which produced the doctrine of *Deo volente*—God willing (I will go). James, speaking to the movers and shakers in the first communities that formed around Jesus's person and message, states, "Pay attention, you who say, 'Today or tomorrow we will go to such-and-such a town. We will stay there a year, buying and selling, and making a profit.' You don't really know about tomorrow. What is your life? You are a mist that appears for only a short while before it vanishes. Here's what you ought to say: 'If the Lord wills, we will live and do this or that.' But now you boast and brag, and all such boasting is evil. It is a sin when someone knows the right thing to do and doesn't do it."

It is unlikely that James, whom tradition sees as the brother of Jesus, was the author of this late first-century text. The Epistle seeks to offer us something more than a pious statement of personal faith or a scholastic doctrine to build on. It was written in a time when the Jews became frustrated with corruption, injustice, and poverty. This led to conflict in the scattered congregations. The earlier chapters of the letter offer practical ways to care for the poor and stand up for the oppressed. Faith or belief, James contends, needs good works of justice. Believers live and do differently. Our text, simple as it might read, brings questions difficult to answer: *How does a person of faith plan a trip without being guilty of James's warning?* Reasonable people do plan, raising doubt that James is against planning. *How does one plan with an openness for God's intervention—those adventus moments we identified? Is all James wants for us to admit is that our plans can go wrong due to divine intervention?* James suggests we need to let go of control and open ourselves to surrender to the experience before us. He reminds us that whatever we do has implications. *Is James against making a profit, even by ethical means? Or if we bring James's ethics into our world, is he only against capitalism and unethical gains?* In the context of poverty in a specific community, James does speak to the wealthy and calls on them to be responsible with their wealth. *Is all that happens the will of God?* James does argue for a close relationship between God and the congregations in need but is not deterministic as he warns against self-assured hubris.

Sustainable Decisions

Caring, as we've discovered, is an obligation a traveler cannot avoid. Rather, despite the challenges caring brings, acts of restoration, restorative justice, and relationships of solidarity are what distinguish a traveler from other persons who are also on the move. Caring, as already alluded to, embraces the mandate for a sustainable future. Sustainability is an inclusive relational concept that assesses the viability of relations between humanity and nature, and it includes matters of social, economic, and heritage justice, gender equality, political participation, and community development. Learning that the richest 10 percent of the world's population—most of us reading this book—are responsible for 49 percent of total lifestyle consumption emissions leads one to admit that this is not just or sustainable.[41] Sustainability addresses much more than CO_2 emissions as we act justly, love mercy, and walk humbly, but it does include our relationship with fossil fuels. We should practice sustainability every day, not just when we leave our homes.

When it comes to sustainability, our mode of transportation and where we stay are important. Walking and cycling are better than driving, rail and bus transportation beats driving nonelectric cars, and airplanes and cruise ships are the worst polluters. Cruise ships, now the size of small cities, have massive diesel engines and produce tons of gray water and waste. The 6,328 pounds of CO_2 emissions our family's flight to Alaska created is a small part of the 4 percent of emissions air travel adds to the annual global emissions output.[42] Sometimes we travel with moral dissonance, as flying is the only reasonable choice. That research indicates travelers are willing to pay a carbon tax to offset their air travel indicates the values travelers have.[43] Similarly, travelers make decisions about their lodging. In today's world, many choices exist, ranging from hotels to family-owned bed-and-breakfast homes, from homestays to home rentals. Each of these lodging options sees money flowing in different directions, from travelers into corporate hands, to landlords, to homeowners. On safari or when island hopping, we seek lodging that pays service personnel a living wage, that supports the education and health care of employees and their families, and that enriches the local community. Even here the decisions we have to make are not easy, as service industry employees are often underpaid, and rental homes and apartments negatively impact the permanent residents around them. When money flows to corporate hands in neoliberal capitalism, the average person is most often exploited and ecological harm follows. Homestays are growing in popularity, especially for single, educated women, because homestays are most often the safest way to stay as they offer local guides to protect, inform, and educate.[44] Sometimes as we choose our mode of transportation or lodging, we choose between the lesser of two evils.

Heritage justice as an aspect of sustainability speaks to protecting historical sites, many of which have sacred meaning to local peoples, as we discovered with Tenam Puente. We also think of Uluru, or Ayers Rock, in central Australia. Only recently was it closed to tourists who bouldered up the monolith despite signs placed by the local Aboriginal Pitjantjatjara people, for many years, stating "Uluru is sacred in our culture, a place of great knowledge. Under our traditional law, climbing is not permitted. This is our home. Please don't climb."[45] As travelers, we make decisions that deepen mutual recognition, care, and solidarity. We partake in the local economy in ways that benefit the local ecology and our hosts, whose self-determination and emancipation we support. We choose meaningful experiences and relationships and catch ourselves when we gaze, exploit, and are entitled.

As we care, we eat and purchase with mindfulness, and should we engage animals, we do so ethically. Consumption is a neoliberal right claimed in questionable freedom. When we put more food on our plate than we eat and add to food waste, when what we consume was produced at great cost to the environment, or when we eat sumptuous buffets while persons living a block away or beyond the harbor fence are malnourished, our caring is inadequate. Likewise, purchasing trinkets and stuff that will be discarded or donated to consignment stores soon after our return discloses our spiritual hunger and insatiable desire. Our care extends to the animals we meet on our travels. *Do we ride elephants in South Asia or Africa or use mules to descend the Grand Canyon—even if those are historic practices—when animals have to be overpowered to do so or when they are exposed to temperatures in which we prefer not to walk? Do we swim with manatees or dolphins, or attract great white sharks with chum for our pleasure?*[46] *Do we hunt when local tribes, such as the Khoisan in Namibia who have hunted for thousands of years, are prohibited from hunting in sustained ways?* Caring in sustainable ways is a mandate in an age of neoliberal commodification of humanity and of nature.

A traveler's code of care is a spiritual practice that rewards us with the knowledge that doing good, protecting and sustaining the natural world, and providing restoration for injustices and inequalities reflect the reign of the Earthroamer. How we travel reflects who we are and what it means to live in a just society. Through our care, we redefine the meaning of "place" as a destination, for "people" and "place" become inseparable. We care to ensure that our children and their children will be able to leave home and be surprised in awe and wonder as they experience distance and difference. We seek to sustain vibrant ecologies to explore. When we care, we leave footprints of another kind—new and deepened relationships, changed lives and communities, healthy ecosystems—and become "good" travelers.[47]

Quality Traveling and the Right Attitudes

In his seminal book *Zen and the Art of Motorcycle Maintenance* (1974), Robert Pirsig reflects on Quality in a fragmented society. He is motorcycle traveling from Minnesota to California with his eleven-year-old son, Chris, and family friends (the Sutherlands). A deep dive into Greek philosophy and long semiautobiographical Chautauqua monologues make *Zen* a challenging, yet captivating read. Pirsig explores Quality—a concept defying definition but pointing to "excellence, worth and goodness"—in technology, rhetoric, and rationality, and he couples it with and compares it to *care*.[48] To care is to have a Zen-like attitude of mindfulness and attentiveness; it is to notice and to remove oneself out of the way. Quality and care are the two inseparable sides of the same coin. The higher value of the two is care, for when the pursuit of Quality coincides with carelessness, inattentiveness, negligence, and a "whatever" mentality, Quality diminishes or is even absent. Pirsig faults "the Church of Reason," as can be found in the academy and scholastic traditions, for seeking rationality void of emotional experience.[49] The lack of a personal presence leads to carelessness. Still, there is no care without the pursuit of Quality, and vice versa, and "a person who cares about what he sees and does is a person who's bound to have some characteristics of Quality."[50]

Caring reflects an "inner peace of mind" similar to a Buddhist monk in meditation or a fisher who casts a line and waits for action. One aspect of care, Pirsig writes, is "just sitting, . . . a meditative practice in which the ideal of a duality of self and object does not dominate one's consciousness."[51] Care is a power that weaves together the caregiver and that which is being cared about. Following Buddhist thinking, Pirsig sees us bracketing our inner tensions and frustrations and becoming present to whatever we do, whether it is welding a mudguard on a motorcycle or caring while traveling. There are no feelings of separateness—us/them, [an object] and me—when we care. With inner peace of mind, we care and can solve problems. *Zen*'s understanding of care is different from our definition of care but does not exclude it either. We defined care as "reaching out to something other than the self; it is neither self-referring nor self-absorbing . . . [and] leading to some type of action."[52] The inner peace of mind we experience informs our actions and reveals Quality as a process and as an end goal.

The differences between a good mechanic and a careless one, or for us between a just traveler and an unprincipled one, are found in how a task or practice is approached and then executed. "[A good mechanic/traveler] has to care," Pirsig writes. "This is an ability about which formal traditional scientific method has nothing to say."[53] We do not care with our minds, but in and through our actions. Mechanics and travelers who care are patient and attentive, have a sense of responsibility, and can even feel burdened by what they notice and experience. Having the right attitude, an inner peace of mind, is all-important. With the right frame of mind, we "produce right actions and right actions produce work which will be a material reflection for others to see of the serenity at the center of it all."[54]

Pirsig's goal for a humanity that balances Quality and care is to "reform the world, and make it a better place to live in."[55] His vision thus overlaps much with our vision of surrendering and being part of systemic change through building caring solidarity. Attempts to bring change through "the Church of Reason"—through abstract ideas and deconstruction—or from a distance inevitably fail. Following *Zen*, how we travel is thus best described in two ways: we bracket our own thoughts and emotions to be present to our experience and to others *and* we enter into relationships of attentiveness, mindfulness, and care as we travel to serve the greater good. We reflect this attitude when we maintain something such as a motorcycle but also when we travel, when we hike up a mountain, when we are just sitting, or when we anticipate a return home.

We travel with an inner peace of mind fully immersed in the traveling experience *and* do so ethically. We are reminded that anticipating, surrendering, meeting, caring, and returning, like "the making of a painting or the fixing of a motorcycle isn't separate from the rest of [our] existence."[56] Being care-less, Pirsig warns, always impacts one's experience in negative ways and leads to frustration and hopelessness, emotions that rob us of "gumption," which the Greeks called "*enthousiasmos*."[57] Gumption, an attitude, supports us in adversity and when we seek answers slow in coming. Changing systemic inequalities takes gumption.

Toward the end of *Zen*, Chris asks his dad,[58]

> "Can I have a motorcycle when I get old enough?"
> "If you take care of it."
> "What do you have to do?"
> "Lots of things. You've been watching me."
> "Will you show me all of them?"
> "Sure."
> "Is it hard?"
> "Not if you have the right attitudes. It's having the right attitudes that's hard."

With the right attitude, Chris is told, one can do much. Traveling is about one's attitude too. Anticipating, surrendering, meeting, and caring are practices that help us find and stay in a traveling attitude that reflects our journey with the Earthroamer. This might be the most challenging part to ensure that just traveling is a Quality experience.

Having reached San Francisco, Pirsig sensed that he may be heading to a mental breakdown, something he experienced before. It was his initial pursuit to define Quality and running into the walls of the "Church of Reason" that led to a hospitalization that greatly impacted Chris. As they have been traveling, they have been untying their (k)nots. It may be projection, but Pirsig noticed the same signs in Chris too. It is then that Chris revisits Pirsig's hospitalization and dreams Pirsig had while on the ride and releases his father from his shame and guilt. "I knew it," Chris repeats as they contemplate the impact of Pirsig's hospitalization.[59] After this untying, Chris asks, "Will I have the right attitudes?"

"I think so. . . . I don't think that will be any problem at all," Pirsig answers.[60] *Zen*, like all traveling, offers moments of confession and redemption, and the book ends with hope: "Trials never end, of course. Unhappiness and misfortune are bound to occur as long as people live, but there is a feeling now, that was not here before, and is not just on the surface of things, but penetrates all the way through: We've won it. It's going to get better now. You can sort of tell these things."[61] Traveling can bring transformation, and trials may never end. Most travels, of course, do end.

Ending with Quality and care in mind, however, is not easy. For it is when we care and find ourselves in relationships such as with family and with friends, French philosopher Jacques Derrida writes, that caring and ending and mourning conflate. "To have a friend," Derrida writes, "to look at him, to follow him with your eyes, to admire him in friendship, is to know in a more intense way, already injured, always insistent, and more and more unforgettable, that one of the two of you will inevitably see the other die. . . . That is the . . . infinitely small tear, which the mourning of friends passes through and endures even before death."[62] Two friends grieve, for they know one will die first. Similarly, when we travel, we know it will end, and returning home is always on our minds and may even remove any peace of mind. We can either detach ourselves from this uncomfortable feeling, diminish the Quality of our friendships, and make the return more bearable or we can seek a Quality experience to the very end.

Chapter 5

RETURNING

Remembering Faithfully and Maintaining Relationships

There *is* a kind of magicness about going far away and then coming back all changed.

—Kate Douglas Smith Wiggin, *New Chronicles of Rebecca*

I carry my roots with me all the time. Rolled up I use them as my pillow.

Mis raíces las cargo siempre conmigo enrollades me sirven de almohada.

—Francisco X. Alarcón, "Roots / Raíces," from *Laughing Tomatoes and Other Spring Poems*

We are told there is no place like home, a truth we learn in infancy. This "no place" brings memories of love and joy and gratitude for some, and painful memories of neglect, violence, and abuse for others. For most persons, home is a combination of many evocative experiences, and home is neither Eden nor the promised land. Even for adults in the homes they have created, there are the daily rhythms that bore, relationships that frustrate, and conflicts that want to tear apart even as they keep together. Of course, home is also the place where we are loved and find belonging, where we care, and where we have intimate relationships and friendships eagerly awaiting our return. We long for the pure joy a family pet displays upon reconnection, we look forward to sleeping on a familiar bed and with our own pillow, and we long for a place where our hands and feet know where we keep stuff and where to land. Part of us always seeks the familiar, because constantly being on the move and being stimulated by everything new are draining. While traveling, much about home moves out of focus. As returning home approaches, however, the joys and concerns of home slowly reenter our minds. We grieve the freedom and hospitality we have received along the way but are about to leave behind: meeting others and the Earthroamer; the time spent with ourselves and in nature. Returning after surrendering, meeting, and caring exposes the ways we and home have changed. Like fault lines in the earth's crust, these transformations—shifts—often cause tremors, landslides, earthquakes, and aftershocks. Research indicates that the farther one travels from home and the longer one is away, the deeper the satisfaction experienced and the more challenging it is to transition back to home.[1]

Some years ago, I undertook a four-week motorcycle trip to Alaska leaving Michigan.[2] The last week was all about heading home. After visiting as many national parks and sites as possible, suddenly little interested me. I was overcome by a homesickness that determined all my actions. I was willing to stay in the saddle longer than before, covering the miles to get home as soon as I could. Part of me was disappointed in myself as the goal of new experiences was replaced by the goal of getting home. As I entered Washington State and turned east, my then five-year-old daughter heard I was heading home. She suddenly developed symptoms that her pediatrician thought might be of emotional origin, as physical causes were ruled out. She too had a form of homesickness because her home, without me, was incomplete, and she longed to have me there. The symptoms disappeared one day after my return when I stepped back into a parental role: playing a game in which she and her sister had to give me back my tickle, which I had lost in Alaska.[3] I discovered that returning from a travel creates its own goals and that reentry has a significant impact on one's life, relationships, and the lives of others.

Sometimes the impact of returning from one's travel lasts longer than a few days. As an academic receiving research leaves, I spent four months in Africa, most of it alone as my family returned to the United States after joining me briefly. The leave was productive, as I finished a book, taught a seminar in Zambia, and lectured in South Africa. For more than three months, I wrote, lectured, met many colleagues, and made new friends. I also spent much time in nature. As an introvert, I enjoyed the gift of solitude. Past returns have brought "traveling blues," and I expected that this time would be no different. After I arrived home, the excitement of being with family again and teaching in familiar classrooms soon faded as a depression-like melancholia beset me. I remained productive, which distinguishes this feeling from depression, but used more alcohol and food than I needed.[4] Returning to the gym and playing tennis, activities I enjoy, caused one injury after the other. My sleeping pattern changed from being an early riser to sleeping in most days. My psyche, soul, and body sent clear messages that I was grieving something, and my familiar surroundings could not replace the losses felt. The reentry process was not going well. The sense of sadness with its melancholic cynicism stayed with me for six months. Whenever I contributed to a conversation, it was a "negative" one, pointing to what lacked, what was "wrong," what change was needed, and what could be "better." One day I noticed I had moved on, but my changed sleeping pattern remained.

Reentry shock, a form of reverse culture shock and a period of readjustment and reintegration, is not uncommon when we return from our travels. Returning's shock brings feelings of sadness, anxiety, depression, and even alienation. Fueled by making comparisons between the places visited and home, and feeling that the benefits of traveling or a holiday are fast disappearing as everyday life returns, reentry shock can be an unsettling experience. The six months of readjustment and cultural confusion I had returning from a research leave were an experience of "friendly loneliness," which Rubem Alves names as the experience of a self being transformed by solitude yet unable to fully comprehend and tell about the change.[5] Friendly loneliness comes from the fantasies that arise in us when we compare ourselves to others, or when we compare our present with our past or anticipated future. Such comparison breeds envy, Alves warns. Though unsettling, he sees potential in friendly loneliness to be a garden in which one can plant flowers. One flower is a changed self.

The Return of the Kaleidoscopic Self

Inevitably, the moment arrives during one's travel when the awareness sets in that the day of return is fast approaching. As we anticipated, planned, and surrendered to our travels, similar emotional, spiritual, and physical preparation is needed to return home. Some will deny this moment until the very end, as it indicates that the time away is about to end. To be enriched by one's traveling experience, however, implies mindfulness even in this last phase of travel. In the first chapter, we noted that as toddlers we practiced rapprochement, which we experienced as an emotional and relational crisis with hesitations and mood swings. We cannot return home after a period away and not remember the hesitant first steps we took to separate ourselves from our parents, only to return to them. Since being mindful about returning feels disloyal toward our time away, we would rather not prepare for the return journey, other than buying the last gifts we want to bring home. This lack of preparation deepens any reentry blues we might experience.

The return trip also beckons us to surrender to the experience, for we can encounter unexpected difficulties: discovering we picked up bedbugs in the last hotel, finding the car battery dead after a light was left on, not being able to find a passport, missing a flight or having one canceled, falling ill as one's vigilance for safe food and water consumption slips in anticipation of returning. Since traveling invigorates but is also tiring, facing adversity is especially challenging on the return trip. When traveling by road, special care is needed, as 90 percent of crash fatalities happen within ten miles from urban areas, which for many would be close to home.[6]

Going "home" needs a brief comment, as the concept of home has emotional, relational, philosophical, and spiritual meanings, among others, each meaning having its own definition. Most often "home" describes that place where one experiences the biggest sense of belonging and selfhood. For many, home includes a physical dwelling, but home can be a language, an identity, a town or city, even a region or country. A dynamic concept, our experience of home changes as we make emotional attachments and find our sense of self supported by persons, communities, and places. Even though we continuously construct and reconstruct our "home," it remains unsettling when we discover our homes changed or at risk.

There is grace in the fact that home is a fluid concept, for the self is fluid too. "[We] are many selves," writes theologian Pamela Cooper-White, "the feeling of unity is an illusion, with adaptational value, but if we have the courage to face our multiplicity, I believe it is a bridge toward empathy for the different, multiple Other."[7] The understanding that a person has multiple selves challenges traditions, secular and religious, that argue for the "unity" of a person or that see persons as a body, a psyche, and a soul. Returning can have one part of a traveler resisting the end of a vacation or trip, with another part looking forward to reconnecting with loved ones, friends, and the familiar, and the third part already engaging the commitments that have accumulated at the office. Becoming torn and confused, or at least scattered, is a reality travelers face on their return journey. We surrendered in our multiplicity to the experience of traveling and met the multiplicity of others and God. When we embrace our multiplicity—that we truly are complex beings with many parts—and recognize the multiplicity in the Earthroamer and in all we meet, new possibilities of self-care but also for renewed and deepening relationships arise.

The belief that the self is made up of multiple selves that present themselves briefly only to move back as if into hiding developed in the late 1960s and early 1970s. Psychologists recognized that the self as an integrated "unit" is more a metaphor than a reality, for the self has intrinsically conflicting and irreconcilable parts. In a classic essay exploring this new awareness, psychoanalyst Jody Messler Davies argues that our multiple selves stem from unconscious fantasy and abstract thought that are never integrated, as well as the internalization of many relationships that are constantly being formed only to lose prominence—a relational unconscious. The multiple self is a dynamic self that was formed in complex relationships.[8] Moments of trauma, deprivation, neglect, grandiosity, and denial greatly strengthen the multiplicity, even to the extent of dissociative disorders. The multiple self is easily recognized in dreams, but these selves are there even in "normal" moments. Should a person dissociate as a defense against trauma and abuse, the hidden selves are revealed. Still, there is nothing pathological about having multiple selves—it is just who we are, much like the Trinitarian God has three persons in one. We are dynamic as past experiences and present states engage memory and the many internalized relationships, each informing a different part of the self. "We [thus] deal not with one unconscious, but with multiple levels of consciousness and unconsciousness—a multiply organized, associationally linked network of meaning attribution and understanding," Davies writes.[9] We are, Davies continues, kaleidoscopic selves, and quoting her extensively on this metaphor can help us move away from thinking in terms of a unitary self:

> I have compared this organization of mind to the working of a child's kaleidoscope: intricate patterns, varied but finite, conflating and reconfiguring themselves from moment to moment. These are patterns whose components will become recognizable over time, yet whose parts will come to light in different constellations of prominence and obfuscation, depending upon their positions relative to each other and to the observer. Those patterns that repeat themselves with regularity—inherent representations and later fantasied elaborations of the most essential formative relationships—come to form a sense of character, order, and agency. They are that which makes us knowable to ourselves and others, despite this kaleidoscopic mélange of multiple selves. They order our world and create, out of chaos, personal templates of constructed meaning.

As a kaleidoscope, we relate to ourselves and others, assign emotion to experiences, and relate to our bodies.[10] Traveling, with all its sensory input, moments of relating, and diverse emotions, is turning our kaleidoscopic self over and over again to create a new and unique configuration of our multiple selves. Returning from our travels is another twist of the kaleidoscope's tube to create a new pattern for the self, holding true to the meaning of "kaleidoscope"—"seeing a beautiful form" (from the Greek *kalos*, meaning "beauty"; *eidos* meaning "form or shape"; and *skopeō* meaning "to look or see").

Blue but Enriched

Research indicates that returning from a vacation impacted by "the divergence between the idealized image of holidays, the expectances posed to balance the daily frustrations ordinary people feel, and the return to work" increases a person's chance of divorce or the ending of a serious relationship.[11] To return from one's travels is risky as the reentry process and reorientation initiate an emotional and relational rollercoaster. We recognize the turmoil in a bad mood, increased anxiety and disorientation, lack of motivation, depression, and a rise in conflict in families and relationships. Psychodynamically, travel blues are tied to the ambivalent loss of someone or something we loved.

Sigmund Freud's classic 1917 essay "Mourning and Melancholia" can help us understand why returning from our travels or even just vacation remains challenging.[12] Freud notes that when we identify strongly with someone or something and they or it disappears—a person dies, we need to leave a newfound destination and friendships, or our vacation ends—we become sad and mourn. We can even lack the capacity to adopt a new object to love. "The loss of a love-object is an excellent opportunity for the ambivalence in love-relationships to make itself effective and come into the open," Freud writes.[13] This ambivalence fuels anger and hostility, and it is thus no surprise that anxiety and conflict increase when one returns home. When the hostility turns inward, however, it manifests as melancholia in one person and depression in another. Whereas Freud saw these dynamics play out especially in child-parent relationships, we can see the lost object as the loss of the freedom, excitement, and wanderlust that are inherent in traveling.

Experiencing sadness during and after returning from traveling highlights the paradoxical nature of leaving home, for traveling to new communities and destinations is mostly associated with positive experiences such as joy, humor, optimism, mindfulness, and savoring.[14] Traveling instills growth in emotional resilience and adaptability due to surrendering to the experience; increased flexibility, assertiveness, and self-confidence; improved problem-solving skills; enhanced perceptual acuity and attentiveness to verbal and nonverbal cues; better communication skills; being more inclusive while also boosting our personal autonomy; and experiencing a shift in values. We become more critical of our own culture while appreciating the cultures we have discovered. The shift in values we experience can also lead to personal confusion, interpersonal miscommunication, and experiences of loneliness, all adding to the stress of returning. Though many of these gains will last but a few weeks upon our return before the grind of daily living erases them, some benefits of travel can last longer.[15] There is nothing like finding a filled inbox and attending nonproductive meetings to remind one of what was lost in returning home. Healthy routines, exercise, a new hobby, relationships, creating a photo album or blog, even planning the next travel can guard the gains of one's travel, if only for a while longer.[16]

Caring, which we said is a way of thinking and doing that maintains, continues, and repairs our world and everything in it, receives new meaning in the context of returning.[17] Mindful that we are impacted by returning just as we are impacted by leaving home and traveling, we practice self-care and other-care. We give ourselves permission to be "off"; we engage those life-giving practices that can sustain us; we are kind to our travel companions who's kaleidoscopic selves were transformed too; those around us who changed while we were away and are being changed by our return; we communicate in such ways that love and trust deepen and distance and conflict are avoided. Caring is especially important as we approach multifaceted (and not multilayered) persons. Each part of us has been transformed by the experiences we have had while traveling as our identities are relationally and culturally constructed. After traveling, we cannot return as the same person who left. Naming how we have been transformed, however, is not easy and may take weeks or months to become clear to us. Initially, we may only feel aftershocks, but when those experiences are seen as invitations to discover new meaning and purpose in who we have become and are becoming, we need not be reactive to this dissonance. Returning transformed does not mean we are a different person. There are personal continuity and definite discontinuity, similar to Jesus on the Emmaus road, when we return home (see chapter 3).

We return changed and possibly with traveler's blues and some melancholia. We also come back enriched. This richness should not be confused with the monetary focus of neoliberal capitalism even if what we return with is a form of capital. French sociologist Pierre Bourdieu explored the role of power in society and used the term *social capital* in his book *Outline of a Theory of Practice* to describe the dynamics behind modes of domination in society and how labor is perceived and institutionalized.[18] He studied how people develop adaptive strategies to engage the structures of their social worlds. We all play the economic game, for example, a game in which most persons lose due to the economic games' self-interest and its inherent disinterest in the well-being of the other. Later, in his essay "Forms of Capital," Bourdieu distinguished between three primary forms of capital: *economic* capital (finance and ownership), *cultural* capital (the arts, sciences, and education), and *social* capital (relationships and status).[19] Bourdieu came to cultural and social capital as he studied the effects of the inequality of education on the poor and noted how cultural and social capital are handed down from one generation to the next in the home. He resisted seeing cultural capital through the lens of economic gains or losses and claimed cultural and social capital have worth. Cultural and social capital—as attempts to counter or deny economic capital—are conversions of economic capital, difficult to describe in monetary terms yet intricately tied to the world of finance. As such, all forms of capital are at risk of privileged irresponsibility.

Traveling increases one's cultural and social capital, even if it negatively impacts one's economic capital in the short term. In an *embodied* state, cultural capital speaks to "long-lasting dispositions of the mind and body."[20] Dispositions speak to gaining a sense of how the game is played, a form of practical wisdom or reasoning about how things work. We "know" intuitively what to do to win or stay in the game of life. There is a personal cost to self-improvement, Bourdieu warns, of investing in oneself. Traveler's blues is one risk. Cultural capital, however, also has an *objectified* state, those cultural goods ranging from media, monuments, museums, science, and forms of technology. Our cultural capital increases as we enter new cultures, visit museums and historical sites, and witness how others live with technology, ranging from Wi-Fi that is not always readily available or lightning fast to different modes of transportation. In its *institutionalized* form, cultural capital is found in academic qualifications.

When we travel, we not only disrupt what has been handed down in the home and our own culture as we allow a different culture to inform and transform us, but we also become rich in ways that deny finance and class-exclusive use of the word *rich*. Rather, we cultivate a new *habitus*, a system of dispositions that function unconsciously akin to a holding environment where one is held by relationships, beliefs, practices, physical nurture, and also nature.

A habitus represents those lasting, incorporated ways of being that inform our doing. It is for social living what muscle memory is for athletes—we can act without having to think about it. Within its habitus, cultural capital also functions as *symbolic* capital, which is unrecognized in the world of finance and the unequal distribution of wealth where the belief is that every person can become rich. Symbolic capital is recognized in cultural worlds where social welfare, expressions of ethnicity, competence, experience, and different abilities not only count but also are encouraged, nurtured, and protected.[21] Symbolic capital, at its best, includes prestige, honor, respect, and positive impressions as it celebrates the uniqueness persons, groups, and cultures bring. This capital, however, turns into *symbolic violence* when structures of dominance are unconsciously incorporated and perpetuated, even by persons resisting the dominance. Symbolic violence indicates a failed habitus; a broken social structure where what is unjust or unequal is ignored, not seen as problematic, or argued for as natural or even just. When we travel, we return with a deeper awareness and appreciation for the cultural capital of others, changing the relationship we have with our own cultural capital.

Traveling also increases one's *social* capital. Social capital is all "the actual or potential resources which are linked to possession of a durable network of more or less institutionalized relationships of mutual acquaintance and recognition—or in other words, to membership in a group."[22] It is one's circle of friends, relationships that now include persons newly met, people who can be called on when needed and who expect the friendship to be maintained and deepened. Social capital also speaks to being part of a group, whether a family, a tribe, or a community that gathers for a specific purpose. Travelers on a mission trip, who do volunteer work during their travels or who remain lengthy periods in one place, build their social capital faster than persons always on the move. Traveling grows our social capital, for it increases the size of our relational network. Whereas neoliberal economic capital seeks to keep its network as small as possible—the top 1 percent, who generally dislikes unions, for example—social capital has the opposite interest. When we visit a collectivist culture during our travels, where a strong sense of community exists, we notice the anemic social capital of our individualistic culture.

Social capital is not a given. It takes continued and conscious work beyond the initial formation of a relationship or group. This alone is reason enough to return to the same destination again, to choose homestays, and to linger in a place. Though economic capital can allow one to travel all over the world and even assist in growing one's cultural capital, such actions may actually prohibit and even diminish one's social capital. When our resources allow us to skim across the earth, we can visit many destinations yet not build any relationships along the way. Furthermore, we may have more time on our hands when we travel, but only because we are buying the time of others, who now have less time for themselves, their families, and others. Bourdieu sees the benefit of social capital as great collaboration and the ability to make a difference, whether that is being a stronger team or being able to address those social issues that need consistent, long-term presence and attention. Relationships of solidarity indicate important social capital. In our social capital relationships and being a member of a group there are constant exchanges taking place, not of money per se but especially of recognition, belonging, support, protection, care, and accountability. Through our social capital, we are endlessly affirmed and reaffirmed. The gift of travel's social capital and exchanging lives leaves one with a "nonspecific indebtedness which is called gratitude."[23]

Traveling *converts* economic capital into cultural, social, and symbolic capital and can be seen as a form of "transnational capital."[24] Returning from traveling is an opportunity to create a mini-habitus, one not spanning society but rather one that includes a traveler's immediate relations. As such, traveling is a form of power similar to other forms of power in that it shapes identities and occupies minds. It enriches us, or in the language of Bourdieu, it instills dispositions and reproduces culture. Just travel is especially important because converted capital always serves economic capital. Surrendering, meeting, and caring undo the potential risks that loom here. So too does the deepening of our social capital when we remember our travels with gratitude.

Remembering with Faithfulness and Gratitude

How we remember our travels can both prolong the positive effects of traveling and diminish the negative experiences we had while traveling or upon reentry. In the previous chapter, Jacques Derrida reminded us that friendship is closely tied to endings and thus mourning. To protect ourselves against the inevitable ending of a relationship—thus loss—we do not fully invest in a relationship. Memory, like relationships, is linked to mourning. To mourn, Derrida writes, is "to reckon, to recount, relate, or narrate, to consider, judge, or evaluate, even to estimate, enumerate, and calculate" the loss that occurred.[25] We can use this understanding of what it is to mourn to inform not only our returning but also our remembering. The period after one has returned from travel is the optimal time "to reckon, to recount, relate, or narrate" our experiences, first to ourselves and then to others. We remember to re-present in the present what is now in the past.[26]

Remembering carries the risk of misrepresentation and betrayal. "Posthumous infidelity," Derrida writes, implies remembering in such a way that one becomes "unfaithful" to the person who died, or for us, toward the travel experiences we have had.[27] We may always be at risk of infidelity as we never fully comprehend even our friends. "There is surely a kind of infidelity in the biography or obituary, which tries to encapsulate a life, to reduce the dead to their accomplishments, to a series of dates and places," Derrida writes.[28] In the rituals around dying, posthumous infidelity is also seen in a person being eulogized and idealized beyond recognition, where bitterness and anger deform the person who died, or where silence denied the person's existence and impact on our lives. Derrida's warning is that no one is all good or all bad and that not engaging the work of mourning is costly.

Our traveling experiences are no different. There are many life-giving moments alongside experiences that surprised in unpleasant or even traumatic ways. Idealizing or demonizing our trips in our portrayals, a tendency easily recognized in many online platforms, will feel inauthentic to the self, for every travel has moments that are confusing, frustrating, frightening, sad, and colorless. In the practices of anticipating, surrendering, meeting, and caring, we inevitably stumble and fail, only to try anew. Likewise, when we diminish or devalue our experiences, a community, or a destination, the disloyalty we portray will carry over into who we are, how we relate, and what we do. And in remaining silent about our travel, whether to protect against shame, guilt, or even disclosure, we do not honor the experiences we've had, the people we've met, and the places we've discovered. When we are guilty of posthumous infidelity in our relationships, we are unable to redeem endings or celebrate the person who died. We can undo our traveling experiences in similar ways as we remember and tell about them.

Doing justice to one's experiences, to remembering faithfully, is difficult. The three infidelities we highlight—of idealization, devaluation, and silence—are always part of the practice of traveling. It is not something that appears only upon our return, for we can be unfaithful in our anticipation, in our surrendering, and in our meeting and caring too. When we remember our travels, we want to remember the uniqueness of what we experienced and do so in our own words, not through the eyes of those who wrote travel brochures or who posted online reviews. We want to situate our experience in larger contexts, for what we experienced does not reflect our destination completely. We tell only after calculating or estimating what really happened, how we influenced what happened, or how we might do things differently the next time. Such reflexivity minimizes the risks of posthumous infidelity. Keeping a travel journal can be of much value here, for we can begin to process and internalize our experiences when what transpired is fresh in our minds. When we remember faithfully, thoughts of "approval, solidarity, and gratitude" are held in tension with disagreements, confusions, and wonderings.[29]

Gratitude may be the widest path to not only avoid risks in remembering but also sustain the gifts of travel in our return and beyond it. Among its many benefits, gratitude keeps us from making comparisons—between our home and the destinations we've visited, for example. "The more comparison there is," writes Danish theologian Søren Kierkegaard, "the more it seems that there is still plenty of time; the more comparison, the more indolent and the more paltry a person's life becomes."[30] When we compare, Kierkegaard warns, we either discover we are far behind others or decide we are ahead of others, neither stance serving our person or our relationships. Reflecting on Jesus's comment that the birds of heaven and the lilies in the field have no worries (Luke 12:22–32), Kierkegaard muses, "The lily is the human being. The naughty little bird is the restless mentality of comparison, which roams far and wide, fitfully and capriciously, and gleans the morbid knowledge of diversity; and just as the bird did not put itself in the lily's place, comparison does the same thing by either putting the human being in someone else's place or putting someone else in his place."[31] Traveling without comparing is impossible, yet comparing rarely facilitates personal growth or deepens one's relationships.

We look at our travel and think others had a better experience. Maybe they had more money to spend and visited destinations we dream about, or they had better weather, or a crisis never came their way. Or in comparing we think the reverse, separating ourselves from others. The Earthroamer is beyond all comparison and calls us to humility and reflexivity—being conscious about our thinking and doing. "Comparison limps [with worry] or is only half truth," we learn; it is damaging because it keeps us focused on a here and now that proves lacking. Rather, Kierkegaard calls us to be content with being a person and to focus on ultimate matters such as walking humbly and justly with the Earthroamer.[32]

Gratitude is the antidote to comparing, especially if one finds one's own reality lacking and envy begins to set in. Psychoanalyst Melanie Klein, who taught us that adult emotions have their roots in infancy, finds envy and gratitude intricately linked.[33] Gratitude, she argues, takes us emotionally to being fed as infants by our caregivers and experiencing deep enjoyment because of that nurture. Gratitude, with its roots in grace, is a powerful emotion that makes us free from resentment and envy. Enjoyment and pleasure, which we experience in our travels, awaken envy and gratitude in one's life. The two emotions are inversely connected. We can experience both at the same time. However, as gratitude increases, envy diminishes, and vice versa. Envy is that "angry feeling that another person possesses and enjoys something desirable—the envious impulse being to take it away or to spoil it."[34] To be grateful is to be aware that we received a gift that possibly cost the giver something. It opens us to "the capacity to love" and to being generous.[35] When we internalize goodness in infancy, the experience becomes the foundation of "hope, trust, and belief in goodness" and expressing gratitude.[36]

When envy sets in, our gratitude diminishes or even disappears, and we have less capacity to be responsive in loving and caring ways to ourselves and others. For Klein, it is no surprise that envy is one of the traditional greatest sins, maybe even the biggest, for it spoils and harms the goodness we experience in life.[37] Envy, with its early roots around the first feedings we received, is a destructive emotion that remains with us as we compare ourselves and find ourselves or our experiences lacking. It attacks the other in the belief that goodness will not return and fuels an insatiable greed that defies need. The envious, greedy person is never satisfied.

Envy and posthumous infidelity are linked through idealization and devaluation, two defenses against envy.[38] These processes can focus either on one's self, on others, or on one's experiences. When envy sets in, we unconsciously "opt" for other attitudes and actions to protect us against envy's internal and external destructiveness: we become ambitious, we become confused as to whether an experience was life-giving or life-depriving, we flee from one relationship to another or become greedy, and we love less and hate more.[39] When envious, our remembering is tinged with nostalgia, which finds our present lacking.

Comparing our society and culture to the societies and cultures we have discovered while traveling easily leads to thinking we are much worse off or much better, depending on the issues compared. Expressions of envy, with its insatiable need for something that will fill the gaps that one is recognizing, will follow. Gratitude, however, awakens a different emotional experience and puts one in a life-giving relationship with persons, communities, and places. Being grateful does not deny adversity. If a pickpocket disrupted one's travels, an accident stopped a road trip, or an act of random violence caused trauma, remembering faithfully will recall and tell about those experiences alongside other life-giving moments received as gifts, such as care received as strangers extending hospitality or a helping hand or as beauty and nature holding us in awe and wonder. Expressing gratitude keeps us from devaluing persons, communities, towns, cities, and destinations and from becoming envious of others who do not seem to have our bad luck. It can also prolong the benefits traveling brought us and initiate a hunger to travel again.[40]

Maintaining One's Object Relationships

As we've internalized our parents, other caregivers, and our experiences in infancy, which greatly influence the ways we anticipate and leave home (see chapter 1), we internalize the destinations and relationships of our travels too. We can "talk" to an internalized parent even after the person has died, for the person now resides in our heads and hearts. From there, our parents continue to have an impact on our lives. When a parent or person was abusive toward us, we can challenge the ways that person now resides within us; we can let go of emotional reactivity to experiences even as we remember those experiences; we can grow into a place where forgiveness is the next step in our personal healing and restoration. Likewise, we can return to a community we've internalized, we can imagine ourselves next to a giant sequoia tree, next to a glacier lake, or in a market square or coffee shop. Psychologists refer to these relationships as object relationships, where an object can be a person, a place, or a thing, even an experience.[41] These relationships are the foundation of object constancy, the ability to keep mental images alive, which makes leaving easier. A definite way to prolong the benefits of travel but also to prepare us for the next moment of leaving home is to maintain the relationships we've established, in person and also by using our imagination and curiosity.

The ways we have to maintain our object relationships are numerous, especially in a technological world. Using technology we curate photo albums or blogs with travel memories; we stay in touch with the persons we've met; we follow the news of the local destination; we join a nonprofit serving a need we discovered while traveling; we deepen our historical and sociopolitical studies of the destination we visited; we have our DNA sequenced to find family connections across time and place; we discover a live webcam that streams images of our favorite destination; we join a "Friends of . . ." international club; we prepare food tasted along the way; we gain new skills in the anticipation of using them during a return trip. Often we tell about our experiences as new awareness comes to us over weeks, months, and even years.

The power of maintaining these relationships is that they are also emotional. Since our emotional lives do not know time, we can keep memories alive and deepen our relationships with persons, communities, and places by literally keeping them in mind. Of course, few will be satisfied with an internalized object relationship only. A natural result of maintaining our travel memories and relationships is a desire to return or travel again. Maybe we decide not to return to a prior destination but rather to return to the region to broaden our cultural, social, and symbolic capital. Nurturing and examining one's object relationships are part of a flourishing life. Our next trip will be different from our previous travels not only because we've changed after leaving home but also because we maintain and deepen our object relationships after our return.

Scripture and Returning Home

The people of God the Earthroamer are always on the move. It is thus no surprise that longing for a home and going home are common themes in Scripture: Adam and Eve were banned from Eden, their lush home, never to return again (Genesis 3); Jacob, we learn, liked to stay home, while Esau hunted (Genesis 25); Rachel, having birthed Joseph with Jacob, begged to return to her homeland (Genesis 30); Israel anticipated a promised land, a new home (Numbers 15); Orpah decided to return home, whereas Ruth decided to follow Naomi into initial homelessness (Ruth 1); the prophet Isaiah sees having a peaceful home as a sign of blessing (Isaiah 32); and losing one's home when taken into exile is a loss to lament (Lamentations 5). Jesus, we learn, entered homes and returned to his hometown (Matthew 13; Luke 4); he often sent people to their homes (Matthew 9; Mark 2); and he tells us his Father's house has many rooms and that he will return there (John 14). We are also reminded that we live in an earthly tent on the way to a more permanent home (2 Corinthians 5).

Other than Jesus telling about his return to the Father, the most famous return story might be the parable of the prodigal or lost son (Luke 15).[42] This seemingly simple story is a complex parable that offers much as we reflect on leaving, surrendering, meeting, caring, and returning home. The parable raises many questions and offers few answers. With tax collectors and sinners gathered around him, Jesus first told of a shepherd who went after one lost sheep and a widow who sought a lost coin, both rejoicing when they found what was lost. He then told of a father with two sons. The younger son came to him and asked for his inheritance because he wanted to travel. The father obliged, giving the son half of everything. In a distant land, the younger son squandered the money. When famine came to that land, he worked for a pig farmer, living among the pigs and eating their food. Recognizing his sinful ways and hopeless situation, he decided to return home. He was yet a long way off when his father saw his son, ran to him, and kissed him. The father clothed him with a nice robe and called the servants to prepare a feast, for a son who was dead was now alive again. The older son, coming off the fields, saw the commotion and became envious, feeling unrecognized. Comparing himself to his younger brother, he told his father that he stayed obediently and worked hard and never got a feast to be celebrated. The father defended his actions in calling for a feast, saying his dead son was alive again. The father's defense in the face of his older son, never going far from control and deprivation, did not convince his son.

The parable is problematic as no father in Jesus's time would honor a son's request to give him his inheritance—his *bios*, which can mean "property" or "life"—while the father was still alive. Rather, that son's asking would be seen as rude and taboo. Was the father just generous, or was he foolish? Did the father know he was being manipulated, or did he turn a blind eye? And if he did know what was going on, why did he go along with his younger son's plans? The father goes beyond what was asked, dividing his worth in half, when the older son would have expected to inherit seven-ninths of the inheritance. Why did the father do that? And if the father knew his younger son's immature personality, he also knew the money would be wasted. What did he give his son in giving him his inheritance? Was giving his younger son his inheritance parental giving or parental depriving? If the latter, the father participated in his son's humiliation as he sat among the pigs. And with regard to the other son, why did the father not respond to his older son's need for recognition? We can raise questions about the younger son too: Was he a disgrace to his father? Did he wish for his father to die? Was he envious of the older, settled brother? He encountered no limits at home and repeats that experience abroad. Was his surrender and decision to return an authentic, deep transformation, or another manipulation? How should we understand the fact that this young man could not even tend or take care of pigs? He does sound pious quoting the pharaoh, who told Moses and Aaron, "I have sinned against the Lord your God and against you" (Exodus 10:16). Is he reacting to feeling guilty? Was he transformed by his travel to return and assume responsibility? What was his reception by his community upon his return, insofar as they saw him humiliating his father?

Besides clothing and food, what restoration or transformation did the son long for and experience? How will the younger son respond to the sage who said, "To travel is better than to arrive"?[43]

The older son brings questions too: Was he wrong in thinking that those who work harder and more faithfully should be paid more? Did he give more than he received? What does he deserve? How painful is it not to be recognized by a parent? Did the father forget his older son? How do significant misperceptions between two people, like the father and older son, happen? How is the distance between them maintained? Do we have paternal failure or fraternal envy, or both? How will the older son learn the excitement of leaving and the risk of returning? What would it take for the older son to have voice? How should we think of persons staying behind, who decide they do not want to leave home? And if we follow tradition, seeing the older father as God, does God act without any foresight into how things are going to unfold? Is God only a God for some, choosing a select few over others? Furthermore, is forgiveness a reparative practice, or does it offend? More questions can be raised around the son who stayed behind but is still searching for a home. Of course, if we see the father as God, as tradition consistently does, and make this a story of exile, restoration, and forgiveness, then we need not bother with these questions. But then we also are not imagining how Jesus's listeners would have heard this story, and we will miss its ironies. Again, no father in Jesus's time would surrender his belongings the ways this father did. Between the passive surrender by the father, the aggressive manipulation by the younger son, and the distant, compliant envy of the older son, we can find some wisdom even if "things are not as they seem; more is going on than meets the eye."[44]

We learn through this parable that the desire to travel or to stay home may not always be "pure" or even made with sober judgment. Both leaving and staying are legitimate choices, and others may support our decisions regardless. We witness how surrendering may or may not be authentic. Future actions by the younger son will affirm whether a true transformation was experienced. The younger son, who met many people, entered into utilitarian relationships of mutuality—his needs were met, and those he hung out with received money, food, and gifts. Neither he nor they discovered themselves or each other anew. With the son's surrendering coming after he met people, we are reminded that the phases of travel as we identified them—surrendering, meeting, and caring—are not sequential phases but spiral moments informing the practice of traveling, with leaving and returning as two capstones. Sitting among pigs and eating their pods, the younger son's internalized relationship with his father remained and may even have grown away from home. It was strong enough to drive him home. We note that even in planning his return, he hopes to change one more time: "I am no longer worthy to be called your son; make me like one of your hired servants" (Luke 15:19). Let me work for pay and be more responsible, he wants to tell his father. The younger son, who first compared himself to his older brother, now compares himself to his father's workers. The older son, of course, is comparing too. The one brother addresses the anxiety by returning home, while the other blames his father. In the absence of any gratitude in the narrative, other than maybe in the father, envy abounds. We want to believe that the younger son has changed as he returned home, but we do not know whether that is true. He left too immature to travel and may very well have returned the same way. What we do see in this story of return is that reentry upsets the existing system and its relationships.

The parable of the prodigal son, as tradition often names this story, is a complex narrative that reminds us that much happens between leaving and returning home, before and after, intrapersonally as well as interpersonally. Some return as if they have never left. Others return profoundly changed by what they experienced. Some experience forgiveness along the way, and others have their painful experiences of rejection and exploitation validated. Like the father in our narrative, those who celebrate with festivities upon our return can either indicate something akin to a father's continual denial of being manipulated by his younger son, or they can be seen as offering us a grateful and grace-filled response of reconnection and restoration. Traveling, like Jesus's words, cannot be reduced to timeless and universal truths. Walking with the Earthroamer remains a mysterious gift.

Returning to Make a Difference

"Tributes made, gifts given, impose a claim on the receiver. A regime of gifts is a regime of debts," writes philosopher Alphonso Lingis while reflecting on traveling.[45] As we return with growing cultural, social, and symbolic capital and experiences to share with others, we are challenged by an important aspect of returning as a practice. This element, which has roots deep in the Judeo-Christian tradition as a significant theme in Scripture, is the Earthroamer flowing blessings to all of humanity and all of creation. This is often depicted as a stream or fountain flowing from God's throne, temple, or house. We first find a river dividing and nurturing landscapes in Eden. The Psalmist picks up on this theme and identifies a river of pure joy giving life to the city (Psalms 36 and 46). The prophets too write about this river. Zechariah envisions "running water [flowing] out from Jerusalem" (Zechariah 14:8 CEB); Ezekiel sees a river six thousand feet (or 1.13 miles) wide, defying crossing. On its banks, trees grow year-round fruit to eat, with leaves that heal and give life (Ezekiel 47:9–12). Joel too sees a fountain flowing from God's house, nurturing valleys (Joel 3:18). Nurturing water that flows from God is such an iconic image that the very last chapter of Scripture—Revelation 22:1–2 (CEB)—tells of "the river of life-giving water, shining like crystal, flowing from the throne of God and the Lamb through the middle of the city's main street." This river too produces fruit to eat and leaves that heal. Sometimes the river or stream texts take on a different form but continue the same tradition.

A man associated with the prophet Elisha died. His widow became bitter, asking: Why, God? Why did he have to die? Why am I on the brink of financial ruin? Who is going to care for my children and me? About to lose her two children to a creditor, she visited with Elisha, angry with him too. Elisha learned that all she had was a small jar of olive oil, used for cooking, for light, and for medicinal purposes. He told her to collect as many empty vessels as possible and then behind closed doors pour her oil into each jar. She followed Elisha's orders and soon all her jars were filled. She returned to Elisha, who told her to sell the oil to her neighbors, use some of the money to pay off her debts, and put the leftover oil to personal use. Soon the blessings the widow received flowed into her neighborhood. Her neighbors were unaware of the source of the oil, yet they benefited greatly from having a precious commodity.

It is a spiritual and ethical mandate to share with others the blessings we have received. That includes the blessings we receive in our travels. The visual image is simple, yet compelling: we are water flowing to persons, communities, and nature. For some, there will be a direct relationship between what they experienced and their volunteering and giving back to their home communities. Some travelers saw the devastating effects of climate change and now clean rivers, protect nature, and campaign against single-use plastic bags. Others will come back renewed and deepen their involvement in their faith communities. Feeling grateful and blessed, they volunteer in their towns and cities. Participating in God's grace as it flows freely to all and over nature takes many forms. When we refuse to be flowing water and dam ourselves up, we become stagnant ponds soon overgrown with algae. Stagnant ponds die. When we flow too fast in judgment or due to a lack of collaboration, we become the disgruntled older son from the parable. Soon we have nothing to share, either because we'll burn out or because we've become corrosive to others or to a cause. If we allow travel's anticipating, surrendering, meeting, and caring to become our teacher, how to partake in our communities will be clear upon our return home.

Roots and Homes in a Postmodern World

The opening poem by Francisco Alarcón celebrates the self, which we have said incorporates many parts. This self has roots that are transportable, just as one takes a favorite pillow on one's travels. Alarcón, a Chicano poet and educator, was born in California but moved back to Mexico from ages six to eighteen. His poem reflects not only his bilingual mind but also his bicultural heritage. Alarcón reframes our understanding of placing down roots. With Alarcón's roots rolled up, however, what location would he call home?

Award-winning Vietnamese filmmaker and Berkeley professor Trinh T. Minh-ha has sentiments that echo Alarcón's. Like Alarcón, she too is "[a wanderer] across language."[46] "Every voyage is the unfolding of a poetic," she writes:[47] "The departure, the cross-over, the fall, the wandering, the discovery, the return, the transformation. If traveling perpetuates a discontinuous state of being, it also satisfies, despite the existential difficulties it often entails, one's insatiable need for detours and displacements in postmodern culture. The complex experience of self and other (the all-other within me and without me) is bound to forms that belong but are subject neither to 'home,' nor to 'abroad.'"

For Minh-ha, traveling implies the redrafting of personal boundaries, a process that continues once one's travels end. This redrafting is a need we bring. Returning is akin to "a process of rehabilitation," one that is especially challenging for those who are dispossessed or who have been forced to leave their home and now search for a new home.[48] In our mobile and postmodern world, home is forever in flux. The postmodern world speaks to many selves, many places, many beliefs and ways of doing (and no universals), many imaginary possibilities, and many movements and travels back and forth. In this world, a phrase such as "I am returning home" may be without meaning, and "home" may only be because we create it over and over again.

Home, we can say as returnees, is that locale where moving and staying overlap. For most of us, that is a spot we can mark on a map, a location best described as "here" and not "there" or "abroad," as "presence" and not "absence" or "displacement," as "inside" and not "outside," as "similar" and not "different," as "full" and not "empty," as "wanted" and not "forced," as "continuity" and not "discontinuity," as "evolving" or "becoming" and not "permanently fixed" or "fluid," as what is "known" and not "unknown." That archetypal stories such as Israel's Exodus and exile revolve around a promised land and a temple lost reflects our need to return home. For most, home is where we have the deepest relationships; where some of what we hope for has been realized; where we can live with generosity and gratitude; it is that place of dwelling after traveling, a place where the magicness of traveling presents itself and rests.[49] We are truly nomadic selves longing for a home.[50]

WE ARE *HOMO ERRABUNDUS*

Walking with the Earthroamer

My toekoms wil reis; my verlede wil rus.
[My future wants to travel; my past wants to rest.]
—Stef Bos, "Stilbaai"

Home and habits and household goods
stayed behind in my village.
I am one behind my back,
and another turned to the sea.
The nape of my neck seethes with good-byes
and my breast with yearning.
—Gabriela Mistral,
"The Immigrant Jew / Emigrada Judía"

"My future wants to travel, my past wants to rest," sings the artist. Leaving, the poet writes, brings loss and ambivalence and exposes our multiplicity. *To travel is to move at the speed of being present to one's experiences, bridging distance and difference through acts of care.* The travel experience begins the moment we anticipate leaving home. Inevitably we are transported to times when anticipation caused excitement and gratitude but also to those moments when we expected with fear and dread. As we leave, our thoughts are filled with promise as we open ourselves to experience even as we are confronted by emotional and relational tensions we have brought along and new experiences that come our way. Some parts of us are hoping to let go of less likable parts or experiences, such as the nagging habits we have but cannot seem to shake, the stress and burnout from work, the relational tensions we are caught up in, and the life-changing decisions that await us. We surrender to whatever the traveling experience brings, for we can trust that the road is sufficient and provides. We remain hopeful and are surprised by the many ways we are held by moments, strangers, spaces, and nature. These moments have liminal qualities as we meet God the Earthroamer, who is behind us, around us, in front of us, and in us. With the heightened sensitivity traveling brings, we notice the unjust and unequal ways history has shaped our world, only to repeat itself for each generation.

We see and meet those we encounter. We enter nature in ways that remind us we were made from topsoil, we are earthlings. Mirroring the Earthroamer, we overcome the distances and differences between people and in our communities. Touched by those we meet and by what we see and experience, we care and enter into transformational relationships of solidarity. Together, we restore our world and a creation straining under the weight of injustices, inequalities, and exploitation. When we return home, we are filled with experiences and gratitude. Intuitively we know that sustaining the relationships we formed with people and places is important. We look at our own community and notice the need for restoration and care around us. Already we anticipate the next moment of leaving home, for the sirens' song calls us to new adventure, exploration, and opportunities to care with others.

Traveling with the Earthroamer

Just Traveling questions traveling habits that are closely tied to visions of American inventiveness, power, and progress, visions that often culminate in claiming to be "the world leader." We certainly did not try to reinvent traveling to make it "better." Nor did we seek to be prescriptive in what anticipating, leaving, surrendering, meeting, caring, and returning might look like. Rather, we celebrate the anthropology of *Homo errabundus*—the roaming or wandering person as celebrated by ancient Greek thought.[1] The book recognizes that there are many ways to get from one place to another, as there is a wide spectrum of attitudes and practices when it comes to traveling. Some attitudes and practices, such as skimming across the globe and mindlessly consuming, differ from attitudes and practices such as caring with others and for the earth and discovering persons in authentic ways. As travelers, we are mindful of where we find ourselves on this spectrum. We engaged select conversation partners, and using other conversation partners will inevitably bring new understandings. We explored traveling with three foci in mind: *In what ways is traveling interwoven with our human nature? How does ethical living and addressing the human condition formed along the fault lines of Orientalism and colonialism inform a practice such as traveling?* and *What potential for faithful living, also away from home, resides in the image of God as the Earthroamer?*

In its original Greek, *metaphor* has connotations of "carrying across" or "transferring." For a mobile world such as ours, where digital technology can transport us across continents within split seconds and the furthest corners of the globe are accessible, God as Earthroamer brings new potential for persons of faith. The metaphor and image, which reframe the image of the God we discover in Scripture in a new way, communicate movement and possibility, presence and mutuality. Theologian Sally McFague highlights "the crucial importance of models in theology. They are comprehensive and at base metaphysical, and a theologian is not able to operate without them. A theologian, to a greater degree than the scientist, does not have the luxury of deciding whether to use models or not, the question is, which models?"[2] Theologians, yes, but all persons of faith need metaphors to reflect on God. At best, these metaphors are inclusive and not gender-biased or patriarchal.

We need a new way to think about God in a mobile world in which inequalities are growing. *What other metaphor or model of God can speak into this world?* Some may argue that God Emmanuel, God with us, is sufficient. However, the metaphor may not be mobile enough, for the Earthroamer moves even when we do not. Others might say God as Spirit might be sufficient, though the Spirit is not grounded in the ways the Earthroamer is. Since no metaphor can capture the full nature of God as experienced by us, God as Earthroamer beckons travelers to discover God in life-affirming and life-informing ways. The image contains elements of seeking, of God roaming the earth in search of something. We can expect that travelers, who do so for many reasons, also carry an element of searching with them.

Possible Selves, Practical Wisdom, and Ego Orgasms

"That which now *is* will soon be *was* even as the *was* was once the *is*," says the therapist.[3] We are reminded that our past, present, and future are connected, with both our past and present experiences belonging to an open future. Traveling takes us into the future from the moment we anticipate leaving to the moment we anticipate returning—the future becomes our present and then our past in special ways when we travel. Conversely, as we anticipate the future, our past and even present experiences are transformed by what is coming our way. Simply put, traveling requires mental agility and reorients our experience of time and, in that process, makes change possible. We've explored many sides to the transformative power of traveling. The kaleidoscopic self, with its multiplicity, challenges the idea of us as unitary selves. It requires diverse forms of transformation as the needs of each part of our selves are different (see chapter 5). Add to this multiplicity the fact that the Earthroamer, through *adventus* moments, breaks into our lives and relationships, and the potential for transformation while traveling is rich. The self's transformation and restoration take us to our possible selves.

Life coaches often work with the concept of one's "possible selves." This is the "ideal self that we would very much like to become. It is also the self we could become and are afraid of becoming."[4] Our possible selves are part of our multiplicity and are hoped-for selves. As such, the possible self is generally optimistic in nature but also includes the dreaded self, the self that fears loneliness, incompetence, being controlled and shamed, or losing control to desire or an addiction. Visualizing our possible selves increases the chance of becoming the person we long to be. It is thus not uncommon for travelers to imagine themselves differently away from home: the introvert becomes more extroverted, the person afraid of heights finds themselves on a tower's lookout deck, the person afraid of intimacy risks vulnerability, the person who saves spends money on a day spa, the person competitive in their career volunteers as an expression of care, and the person tied in (k)nots finds release and new ways of being. Allowing our possible selves to take form is a kind of rebirth.

Brazilian novelist Paulo Coelho knows how others can step on one's possible self. As a teenager, he wanted to become a writer, only to be told by his mother that his engineer father would never allow that. This tension ended with his being placed in a mental institution when he was seventeen, a place from which he repeatedly escaped. Three years later, he was released and became a vagabond, traveling for extensive periods of time through Mexico, South America, Europe, and North Africa. On his return to Brazil, he became a songwriter, journalist, and actor, and he was incarcerated by the military government for his subversive activities. Married with two small children, he was thirty-nine but tied in various (k)nots. It was then, in 1986, that he walked the five-hundred-mile *Camino de Santiago* (The Road of Santiago de Compostela) in northern Spain and had a spiritual awakening. This experience became a book, *The Pilgrimage*, which along with his next book, *The Alchemist*, cemented Coelho's status as a writer. In *The Pilgrimage*, Coelho writes, "When you travel, you experience in a very practical way, the act of rebirth. You confront completely new situations, the day passes more slowly, and on most journeys you don't even understand the language the people speak. So you are like a child just out of the womb. You begin to attach much more importance to the things around you because your survival depends upon them. You begin to be more accessible to others because they may be able to help you in different situations."[5]

This "birth" opens one to delight and beauty and makes one feel truly alive. Coelho notes powerful dynamics and emotions when we travel: time slows down, we feel like a stranger, we regress in our emotional age, we are vulnerable, we receive what the road provides as a gift, relationships deepen, we build lifelong memories, and we feel fully alive. He is correct: traveling, which we said does not depend on distance or duration, promises a transformational experience akin to experiencing a rebirth. Whether one travels in one's mind or across town or leaves home for a few days or weeks, the promise remains the same.

What makes the birth of our possible selves so powerful is, in part, that it is an unspoken part of who we are. It is not as if we go around to all we know and say, *I'm going to travel to be more of an extrovert*, or *I'm traveling to overcome my dreaded self and some old habits that are gaining strength*, or *When I return, I want to be different as a partner, as a parent, as a professional. . . .* No, the hopes we have for our restoration and rebirth are most often deeply personal and private. Our possible selves are secret, and we may share them only with a person we meet on the road but not with those nearest and dearest to us.

Just as we may lack the language to speak about our possible selves, so too it is difficult to describe the transformation or rebirth we have experienced from traveling. Hindsight and time bring wisdom, but traveling, with its liminal qualities, often refuses to reveal its full impact on our lives. There is a Sanskrit word that describes this key trait of traveling, *anirvacanīya*, which means "that which cannot be explicated."[6] Traveling's liminal qualities ensure that our minds cannot fully understand what we are experiencing or have experienced, and our language accordingly fails to express to others what happened to us after we left home. One general way to describe the impact of traveling, however, is that one gains practical wisdom.

The ancient Greeks valued *phronesis*, a form of sagely wisdom.[7] *Phronesis* reflects a decision-making process that has practical ends. It discerns good goals to pursue and how best to pursue those goals. For travelers, every phase of the traveling journey—anticipating, surrendering, meeting, caring, and returning—demands practical wisdom and highlights the importance of how things are done. Since every traveler has a personal way of leaving, for example, how they do so is informed by their practical wisdom. The more experience one has of leaving, the more adept one becomes in doing so. The act of traveling thus builds our practical wisdom even as we use it to make decisions while traveling and become better at traveling. Since our practical wisdom is also informed by our relationships with the Earthroamer and those we care with in solidarity, the nature and depth of our practical wisdom increase exponentially when we travel. In the previous chapter, we identified some benefits gained from traveling in terms of social and cultural capital, even symbolic capital, and these "capitals" are important, but growing in *phronesis* might be the most important gift traveling provides. Frankly, without practical wisdom, traveling can unravel into problematic situations that will erode the gifts of travel. It can also put one in physical danger. *Phronesis*, the Greeks believed, leads to the art of living, recognized in happiness (*eudaimonia)* and the emotional and spiritual highs we experience while traveling.

Traveling provides us with many peak experiences, those rich moments that reflect the joy, happiness, and satisfaction of a life well lived. These are moments when our hoped-for selves shine. These experiences of intensified awareness become high points in our lives. The best travel memories never really end, as they stay with us in profound ways. These memories encode themselves in our minds, sustaining us long into the future and until the next moment of traveling arrives. We can call these experiences ego orgasms, "a highly satisfactory experience such as may be obtained at a concert or at the theatre or in a friendship."[8] These experiences often show when our defenses have relinquished their control, which already started when we anticipated our travels and quickened when we surrendered and met others in authentic ways. Ego orgasm identifies the satisfaction of creative living deeper than normal enjoyment or happiness. It is akin to, but goes beyond, sexual climax, as one's whole body is involved.

Traveling, British psychoanalyst Masud Kahn writes, induces ego orgasms, as the experience often includes friendships and surrendering. He tells of a man who traveled with another family to Morocco. There they had wonderful experiences and did a lot of things together. The man felt happy about the trip and "the experience had saturated itself to completion."[9] We describe ego orgasms using words and phrases such as "great," "I have no words," "wonderful," "splendid," "relaxed," "enjoyed," "worthwhile," and the like. In our day-to-day lives, we rarely experience ego orgasms because the grind of our lives erodes moments of emotional, relational, spiritual, and experiential bliss. Immersing ourselves in aesthetics, the arts, religious practices or meditation, playing, and even activities that involve risk can facilitate moments of ego orgasm. Traveling facilitates the same experience as we stand in awe and wonder before the splendor of nature; when we stand on ancient city walls and imagine past battles fought; as we finally reach the pinnacle, having climbed since before dawn; after successfully completing a difficult day; as we look with others at the work completed and recognize that what has been built will change lives; when we meet someone and talk into the early morning hours. In these and similar moments, we feel fully alive, a sure indication of an ego orgasm.

Traveling Has a Future, Now and Tomorrow

Reading a book such as *I Really Should Have Stayed Home: The Worst Journeys from Harare to Eternity* may dissuade some from traveling.[10] Yet the traveling spirit is resilient. We travel for many reasons, including to manifest our possible selves, to gain practical wisdom, and to experience ego orgasms. Since we have been created in the image of the Earthroamer, traveling is and will remain part of the human experience regardless of humanity's plight. Political and social divisions, discord, and xenophobia as well as wars and conflicts cannot snuff out traveling's flame. Pandemics, we've learned, can certainly throw a spanner into traveling's works, and international travel may never be the same again. But when international travel is challenging, we see the rentals of motor homes increase as people embrace the open road. The travel bug's itch is ignorant of geopolitical matters. For most of us, a trip is but a few clicks on our computers or a few taps on our phones away from becoming a reality.

We have sleepwalked into new ways of traveling, and that trend will continue. We have become accustomed to pack only small quantities of fluids when we fly. Our identities have been digitized, and we find ourselves under constant surveillance. Airplane seats have become smaller and aisles narrower as more people are squeezed into a plane. We have grown impatient with the fees attached to and the security scans of checked luggage and now travel with onboard luggage only. We dutifully take off our shoes and belts. We follow news reports where religious and political radicals attack tourists but are unmoved as we plan our next trip. Immunization book and mask in hand, we'll seek those borders that will allow us in or go on that long-anticipated cross-country trip. Between maintaining physical health and financial stability in the world, a renewed sense of vulnerability is with us.

We discovered borders are porous, that travelers can spread disease as explorers-turned-colonizers did before, and that airports and planes can be ghost spaces. We saw that when it comes to health care, the poor are affected more than the rich, persons of color more than whites. Suddenly, attending a funeral may not be reason enough to travel. We wondered about governments that overplay and underplay their responsibilities toward their people. We witnessed narcissism that puts self-interest above the well-being of the community. We have seen solidarity in communities that sing to one another and to honor first responders, communities in which neighbors walk with a Black man afraid to walk by himself. We have been reminded about our own embodiment when we cough or sneeze and others look at us as if we have a contagious disease. We have renewed awareness of the heavy toll of social isolation. Our sense of interdependence has increased. We saw resilience in young and old and a shared humanity of untold people in countless cities as we name racial and other inequalities. Despite ever-changing landscapes, some more hopeful than others, our roaming spirit remains and will find ways to travel.

No doubt our future traveling will look very different. Many will favor national travel above international travel and cruises. Others will drive and not fly. Some will avoid large cities in favor of remote destinations; they will collaborate with hosts rather than stay in a hotel. Technology too will bring new possibilities. Already we enjoy webcams and video safaris, looking at our favorite destinations in real time. Augmented reality—providing an interactive, sensory-rich experience in which our reality is enhanced by computer-generated visuals and information—will give us an authentic experience without having to leave our home.[11] For older persons or persons with different abilities unable to physically leave their homes to go on a trip, augmented reality brings much promise. Regardless of how we travel in years to come, the phases of travel—anticipating, surrendering, meeting, caring, and returning—not only remain pertinent but also hold the promises of rich experiences and of restoration, as we have explored.

We Travel with the Earthroamer's Blessings

Those following the Earthroamer and leaving their homes are never far removed from rich blessings as *adventus* moments consistently surprise on the road. It's fitting that we conclude *Just Traveling* with three blessings for all in need of traveling mercies. The first is a foundational blessing we receive with our spiritual ancestor Abram when the Earthroamer told him (Genesis 17:1, 7, 8), "I am God Almighty; walk before me faithfully and be blameless. . . . I will establish my covenant as an everlasting covenant between me and you and your descendants after you for the generations to come, to be your God and the God of your descendants after you. . . . I will be their God." The Earthroamer seeks relationship with us covers us in blessing. The irony that Abram was yet without any offspring when he received this blessing reminds us that God provides and that the blessings we receive do not depend on our own doing. Rather, this blessing reveals a radical faith, which Abram promptly failed when he laughed at the thought of an old man of ninety-nine having a son. Yet grace gave Abram another chance, and soon Sarah was pregnant, and their lives were filled with possibility. We can travel with expectation and in hope, for we are in the generations recognized by the Earthroamer, who is behind us, around us, in front of us, and in us.

Whereas Abram's blessing originated around liturgical practices, the next blessing probably came from a tavern—a bar—close to a harbor. This well-known Irish blessing, which land travelers took over from sailors (or those who lifted a cup in the tavern), is optimistic if not idealistic. Still, travelers know that there is always another road, that the headwind will change direction, that clouds will disappear, and that the road brings possibility: "May the road rise to meet you. May the wind be always at your back. May the sun shine warm upon your face, the rains fall soft upon your fields." Although this blessing adorns countless Hallmark cards and picture frames, travelers who know they need to surrender, meet, and care also need a blessing that speaks to the injustices, inequalities, and exploitations of this world. *Just Traveling* celebrates the disruptive nature of traveling and offers a blessing to all seeking to travel differently:[12]

> *And now, may the Earthroamer who is behind you, around you, in front of you, and in you also torment you.*
>
> *May God keep before you the faces of the hungry, the lonely, the rejected, and the despised, also the beauty of nature, hospitality, and the kindness of strangers.*
>
> *May the Lord afflict you as you meet and witness the hurt, the wounded, the oppressed, the abused, the victims of violence. May you be reminded that you are part of humanity and nature as you witness the exploitation and destruction of communities and the earth's natural resources. May the Earthroamer grace you*

> *with agony, a burning thirst for humility, justice, righteousness, equality, and solidarity.*
>
> *May the Earthroamer give you courage and strength and compassion to make ours a better world, to make your community a better community. May you do your best to make it so, whether at home or when you are away from home, and after you have done your best, may the Earthroamer grant you peace.*

We receive these blessings with gratitude, as we are *Homo errabundus*—through traveling, we deepen our humanity, and a part of us, lost in this world, returns to us.

NOTES

Preface

1 Adam Phillips, *Monogamy* (New York: Pantheon, 1996), 56.

Traveling: A Paradoxical Practice

1 "Highway 2" appears on Drew Nelson, *Dusty Road to Beulah Land*, Waterbug Records, 2009.

2 Camille Paglia, *Sexual Personae: Art and Decadence from Nefertiti to Emily Dickinson* (New Haven, CT: Yale University Press, 1990), 258.

3 Friend and fellow author, spiritual director, and academic Steven Chase often says one needs to live by the speed of prayer.

4 Tazim Jamal, *Justice and Ethics in Tourism*, 1st ed. (London: Routledge, 2019), 13.

5 Marielle Risse, "White Knee Socks versus Photojournalist Vests: Distinguishing between Travelers and Tourists," in *Travel Culture: Essays on What Makes Us Go*, ed. Carol T. Williams (Westport, CT: Praeger, 1998), 41–42. Risse is an American expat teaching in Oman.

6 I first explored the spectrum that defines both traveling and tourism in Jaco J. Hamman, *A Play-Full Life: Slowing Down and Seeking Peace* (Cleveland, OH: Pilgrim, 2011), 79–80.

7 Chris de Burgh, "A Spaceman Came Travelling," *Spanish Train and Other Stories*, A&M Records, 1975.

8 Maya Angelou, *All God's Children Need Traveling Shoes*, 1st Random House ed. (New York: Random House, 1986).

9 Adam Phillips, *Equals* (London: Faber & Faber, 2002), 122. Phillips writes "On What We Need" to discuss essentialism in psychoanalysis.

10 Karen Baker-Fletcher, "God with Us in the Dust," in *Constructive Theology: A Contemporary Approach to Classical Themes*, ed. Serene Jones and Paul Lakeland (Minneapolis: Fortress, 2005), 180.

11 Abraham Joshua Heschel, *The Prophets*, 1st Perennial Classics ed. (New York: Harper Perennial, 2001), 242; Abraham Joshua Heschel, *God in Search of Man: A Philosophy of Judaism* (New York: Farrar, 1955), 239.
12 Rubem Alves, *The Best Chronicles of Rubem Alves*, trans. Glenn A. Cheney (Hanover, CT: New London Librarium, 2017), 29.
13 Arnold van Gennep, *The Rites of Passage* (London: Routledge & Kegan Paul, 1960), 10.
14 van Gennep, 10.
15 van Gennep, 11.
16 van Gennep, 189.
17 van Gennep, 16.
18 Victor Witter Turner, *The Ritual Process: Structure and Anti-structure* (London: Routledge & Kegan Paul, 1969), 131.
19 Phyllis Trible, *Texts of Terror: Literary-Feminist Readings of Biblical Narratives*, vol. 13, Overtures to Biblical Theology (Philadelphia: Fortress, 1984), 28. For an exploration of race, ethnicity, color, and identity around the figure of Hagar and how she is received especially by African Americans and got linked to being Black or Blackness, see Nyasha Junior, *Reimagining Hagar: Blackness and Bible* (New York: Oxford University Press, 2019). Junior warns against viewing biblical figures as the normative white as she explores "how, when, and why some interpreters choose to identify the Biblical Hagar as a Black woman" (13).
20 van Gennep, *Rites of Passage*, 191.
21 Cultural theorist Edward Said, in his classic 1982 essay "Traveling Theory," identifies four phases to both travel and the evolution of new ideas: departure from one's/a point of origin; passage through different contexts; transplantation into a new context/identity, demanding new conditions of acceptance; and reemergence of the initial identity/idea, transformed by its displacement and new uses. See Edward W. Said, "Traveling Theory," in *World Literature in Theory*, ed. David Damrosch (New York: Wiley-Blackwell, 2014), 114–33.
22 For a call for a theology from below, see Dietrich Bonhoeffer, *Letters and Papers from Prison* (New York: Macmillan, 1972), 17.
23 Alain de Botton, *The Art of Travel* (London: Hamish Hamilton, 2002), 9.

Chapter 1: Anticipating

1 Nicholas Samstag, "Review of Going Sane: Maps of Happiness by Adam Phillips," *Psychotherapy* 43, no. 3 (2006): 363.
2 Samstag, 363.
3 This section draws on the work of developmental psychologist Erik Erikson. See Erik H. Erikson, *Identity, Youth, and Crisis* (New York: W. W. Norton, 1968), 94.
4 This section draws on the work of German philosopher and phenomenologist Edmund Husserl. Phenomenology seeks to make rational the deeply personal and subjective as it reflects on experience, things in themselves, and our consciousness of an experience. Husserl saw anticipation as precognition (*Vorbekanntheit).* He traveled often, not only in Europe, but also between mathematics and philosophy, between the human and the transcendent(al), and between Judaism and

Christianity. See Edmund Husserl, *Experience and Judgment: Investigations in a Genealogy of Logic*, trans. Ludwig Landgrebe (London: Routledge, 1973), 31.

5 Husserl, 32.

6 For a discussion of Husserl, expectation, and anticipation, see Rebecca Bryant and Daniel M. Knight, *The Anthropology of the Future* (Cambridge: Cambridge University Press, 2019), 22.

7 Bryant and Knight, 27. See also Neal DeRoo, *Futurity in Phenomenology: Promise and Method in Husserl, Levinas, and Derrida* (New York: Fordham University Press, 2013), 44.

8 Philip L. Pearce, *Tourist Behaviour and the Contemporary World* (Buffalo, NY: Channel View, 2011), 109.

9 This section draws on Rahner's concept of *Vorgriff*, an imaginative act of knowing not unlike anticipation's empty reach for the future. See Karl Rahner, *Spirit in the World* (New York: Continuum, 1994), 208.

10 Karl Rahner, *Foundations of Christian Faith: An Introduction to the Idea of Christianity* (New York: Seabury, 1978), 34.

11 Winfred George Phillips, "Rahner's Transcendental Deduction of the Vorgriff," *Thomist: A Speculative Quarterly Review* 56, no. 2 (1992): 263.

12 Time as *futurum* and as *adventus* is the contribution of theologian Jürgen Moltmann. See Jürgen Moltmann, *The Coming of God: Christian Eschatology* (London: SCM, 1996), 25.

13 Sarah Blaffer Hrdy, *Mother Nature: A History of Mothers, Infants, and Natural Selection*, 1st ed. (New York: Pantheon, 1999), 501.

14 Our interest is not in the infant–primary caregiver relationship per se but rather in the fact that there is no leaving for traveling without our emotional ties being deeply affected. Attachment theory, as the psychology that explores our emotional ties, was proposed by post–World War II British theoreticians and clinicians such as Margaret Mahler, John Bowlby, and Mary Ainsworth. Attachment theory no longer holds on to a narrow focus on the mother-infant relationship but rather sees a rich facilitative environment raising a child. See Margaret S. Mahler, Fred Pine, and Anni Bergman, *The Psychological Birth of the Human Infant: Symbiosis and Individuation* (London: Hutchinson, 1975), 53.

15 Jessica Benjamin, *The Bonds of Love: Psychoanalysis, Feminism, and the Problem of Domination*, 1st ed. (New York: Pantheon, 1988), 140.

16 Robert M. Pirsig, *Zen and the Art of Motorcycle Maintenance: An Inquiry into Values*, Quill ed. (New York: Harper Collins, 2009); Anne Lamott, *Traveling Mercies: Some Thoughts on Faith*, 1st ed. (New York: Pantheon, 1999); John R. Stilgoe, *Outside Lies Magic: Regaining History and Awareness in Everyday Places* (New York: Walker, 1998).

17 Jean-Yves Lacoste, "The Phenomenality of Anticipation," in *Phenomenology and Eschatology: Not Yet in the Now*, ed. Neal DeRoo and John Panteleimon Manoussakis (Farnham: Ashgate, 2009), 15. Philosopher Jean-Yves Lacoste distinguishes between anticipation's "pre-experience"—projecting past experience into the future—and its "pre-givenness, [that which is] given to us inchoately, in the mode of a hint or promise." Pre-experience and pre-givenness are the Dutch *voorpret*.

18 Angelou, *All God's Children*, 4–5.

19 Angelou, 12.

Chapter 2: Surrendering

1 Barbara Pizer, "When the Crunch Is a (K)not: A Crimp in Relational Dialogue," *Psychoanalytic Dialogues* 13, no. 2 (2003): 174.

2 This is the opening line of Psalm 121: "I lift up my eyes to the mountains—where does my help come from?"

3 Emmanuel Ghent, "Masochism, Submission, Surrender: Masochism as a Perversion of Surrender," *Contemporary Psychoanalysis* 26, no. 1 (1990): 108–36. Ghent draws on D. W. Winnicott's concepts of the true and false self and the fear of breakdown. See also Jessica Benjamin, "Surrender and Self-Discovery: Ghent's Reflections on the 'Force for Growth' and What Impedes It," in *The Collected Papers of Emmanuel Ghent* (New York: Routledge, 2018), 71–76.

4 For *amae*, see the works of Japanese analyst Takeo Doi, *The Anatomy of Dependence* (Tokyo: Kodansha International, 1973); and Takeo Doi, *The Anatomy of Self: The Individual versus Society* (Tokyo: Kodansha International, 1986).

5 Jessica Benjamin, "Two-Way Streets: Recognition of Difference and the Intersubjective Third," *Differences* 17, no. 1 (2006): 122–23.

6 For digital technology creating an e-third, see Jaco J. Hamman, *Growing Down: Theology and Human Nature in the Virtual Age* (Waco, TX: Baylor University Press, 2017), 152–57.

7 For a discussion of these defenses, see Albert Ellis and Windy Dryden, *The Practice of Rational Emotive Behavior Therapy*, 2nd ed. (New York: Springer, 1997).

8 For faith deeper than religion, see Michael Eigen, "The Area of Faith in Winnicott, Lacan and Bion," *International Journal of Psycho-Analysis* 62 (1981): 413–33.

9 For explorations of atonement and sacrifice beyond traditional interpretations, see David V. Meconi, *On Self-Harm, Narcissism, Atonement, and the Vulnerable Christ* (New York: Bloomsbury Academic, 2019); A. M. Gray, *Charity in Rabbinic Judaism: Atonement, Rewards, and Righteousness* (New York: Routledge, 2019); Stephen Burnhope, *Atonement and the New Perspective: The God of Israel, Covenant, and the Cross* (Eugene, OR: Pickwick, 2018); and Anna Fisk, *Sex, Sin, and Our Selves: Encounters in Feminist Theology and Contemporary Women's Literature* (Eugene, OR: Wipf & Stock, 2014).

10 Leviticus 19:18 says, "Do not seek revenge or bear a grudge against anyone among your people, but love your neighbor as yourself. I am the Lord."

11 For a rich discussion of this and other parables, see Amy-Jill Levine, *Short Stories by Jesus: The Enigmatic Parables of a Controversial Rabbi* (Nashville: Abingdon, 2014).

12 Here I use the translation of Eugene Peterson's *The Message*, edited to be gender neutral. I highlight chapter 5, pages 22–24. See Eugene H. Peterson, *The Message: The Bible in Contemporary Language* (Colorado Springs: NavPress, 2002).

13 Commentators highlight Paul's use of Isaiah 32:15–18, 57:15–16, and 61:11–12; and Joel 2:28–32.

14 Len Sperry and Jonathan J. Sperry, "Psychotherapy and Virtue: Positive Psychology and Spiritual Well-Being," in *Religion, Spirituality, and Positive Psychology: Understanding the Psychological Fruits of Faith*, ed. Thomas G. Plante (Santa Barbara, CA: Praeger, 2012), 149.

15 Mary Stewart Van Leeuwen, *My Brother's Keeper: What the Social Sciences Do (and Don't) Tell Us about Masculinity* (Downers Grove, IL: InterVarsity, 2002), 32.

16 *Oxford English Dictionary (Online)*, s.v. "grace," accessed May 6, 2020, https://www.oed.com.
17 A number of Dutch theologians picked up Calvin's sparse references in his *Institutes of the Christian Religion* to common grace and expanded his ideas. See Abraham Kuyper, *De Gemeene Gratie in Wetenschap En Kunst*, vols. 1–3 (Leiden, Netherlands: Höveker & Wormser, 1905); Louis Berkhof, *Systematic Theology* (Grand Rapids, MI: W. B. Eerdmans, 1996); and Herman Johan Bavinck, John Bolt, and John Vriend, *Reformed Dogmatics*, 4 vols. (Grand Rapids, MI: Baker Academic, 2003). Kuyper's *De Gemeene Gratie* translates as "the common grace." He urged Christians to be active in the public domain of common grace as Kuyper pursued the value of non-Christian culture, science, and philosophy. Bavinck viewed common grace more positively than Kuyper did and called it a *metabasis eis allo genos* (a shift to another category or genus) to differentiate common grace from special grace.
18 Berkhof, *Systematic Theology*, 222.
19 Abraham Kuyper, *De Gemeene Gratie in Wetenschap En Kunst*, vol. 2 (Leiden, Netherlands: Höveker & Wormser, 1905), 29. In *De Gemeene Gratie*, Kuyper, a converted humanist, remains ambivalent about the ways particular grace and common grace relate and how to evaluate non-Christian thought.
20 Sytse U. Zuidema, *Communication and Confrontation: A Philosophical Appraisal and Critique of Modern Society and Contemporary Thought* (Kampen, Netherlands: Kok, 1972), 2.
21 Martin Buber, *I and Thou* (New York: Charles Scribner's Sons, 1958), 11–12.
22 Christopher Bollas, *The Shadow of the Object: Psychoanalysis of the Unthought Known* (London: Free Association, 1987), 14.
23 Bollas, 14.

Chapter 3: Meeting

1 Binyavanga Wainaina, "How to Write about Africa," *Granta* 92, The View from Africa (2005): 93.
2 Edward W. Said, *Orientalism*, 1st ed. (New York: Pantheon, 1978), 259. Said has been described as a liminal person: Arab but not Muslim, Palestinian yet American, a scholar and activist, a naturalized migrant, a person who recognizes the individual within larger power structures, and a humanist critical of humanism.
3 For the importance of *location* in postcolonial studies, see Homi K. Bhabha, *The Location of Culture* (New York: Routledge, 1994).
4 The Swedish naturalist Carl Linnaeus (1707–78), known as "the father of taxonomy," created classification systems that coded difference, whether in plants, minerals, animals, or peoples. Though Linnaeus did not classify persons by race, cultural adaptation of his theories did just that, seeing two of the same kind—two humans, for example—as distinctively different. Linnaeus's thought built Eurocentrism, as did the thought of Charles Darwin (1809–82), who gave us evolutionary theory. Though Darwin wrote with compassion about tribes he encountered on his travels and his family was against slavery, his evolutionary thought also led European culture to see some forms of life, including persons, as more evolved than others.

In later editions of *The Origins of Species* and also in his *The Power of Place*, Darwin reduced the importance of natural selection and showed how many other factors influence human formation. Christianity's missionary zeal is a third power that coincided with imperialism and colonialism. Religion and doctrine sought to bring light and good news to dark-skinned peoples living on "dark" continents.

5 David Spurr, *The Rhetoric of Empire: Colonial Discourse in Journalism, Travel Writing, and Imperial Administration* (Durham, NC: Duke University Press, 1993).

6 For an analysis of the dynamics the colonizer used, see Frantz Fanon, *The Wretched of the Earth* (New York: Grove, 1963). See also Ashis Nandy, *The Intimate Enemy: Loss and Recovery of Self under Colonialism* (New Delhi: Oxford University Press, 1988).

7 Modern politicians who claim to be "strong" leaders use the same divide-and-conquer model of leadership to equally devastating effect.

8 Fanon, *Wretched of the Earth*, 18. Similarly, philosopher and Columbia University professor Gayatri Spivak famously asked, "Can the subaltern speak?" expanding the terms introduced by the Italian Marxist philosopher Antonio Gramsci. The subaltern is the person socially, politically, and geographically outside the hierarchy and the cultural hegemony of the colonial power structure. See Gayatri Chakravorty Spivak, *Can the Subaltern Speak?* (Basingstoke, UK: Macmillan, 1988).

9 Benhabib describes elements of relational *intelligence*. Seyla Benhabib, *Situating the Self: Gender, Community, and Postmodernism in Contemporary Ethics* (New York: Routledge, 1992), 52.

10 Hamman, *Growing Down*, 52.

11 Mirror neurons complicate human development, since they are selective in what they mirror. They play a role in what is often referred to as "the automaticity of being" and "the Chameleon Effect," where mirroring takes place blindly or unconsciously. For a discussion of recent research on mirror neurons, see Ruth Leys, *Newborn Imitation: The Stakes of a Controversy*, Cambridge Elements: Histories of Emotions and the Senses, ed. Jan Plamper (Cambridge: Cambridge University Press, 2020).

12 Benjamin, *Bonds of Love*, 15–16.

13 Cecilia de Bernardi, "A Critical Realist Appraisal of Authenticity in Tourism: The Case of the Sámi," *Journal of Critical Realism* 18, no. 4 (2019): 438.

14 Bhabha, *Location of Culture*, 110.

15 Melanie Tervalon and Jann Murray-Garcia, "Cultural Humility versus Cultural Competence: A Critical Distinction in Defining Physician Training Outcomes in Multicultural Education," *Journal of Health Care for the Poor and Underserved* 9, no. 2 (1998): 117.

16 Tervalon and Murray-Garcia, 119.

17 Gloria E. Anzaldúa et al., "Speaking across the Divide," *Studies in American Indian Literatures* 15, no. 3/4 (2003): 10. Anzaldúa also introduced the concept of *mestizaje*, a state of being beyond binary (either/or) conceptions seen especially in Chicanas, women of color, and queer communities. *Mestiza* are marked by feelings of insecurity and psychic restlessness. Anzaldúa called for a "new *mestiza*" consciousness that embraces ambiguity, hybrid identities, and racial mixing, a neutral identity refusing to take sides, opening *mestiza* to all persons. It speaks of a hybrid identity that joins people together, which is different from capitalist and postcolonial hybrid identities that also tie people together. There is potential for *mestiza* to

inform the identity of the traveler. See Gloria Anzaldúa, *Borderlands / La Frontera: The New Mestiza* (San Francisco: Aunt Lute, 1987), 5.

18 For a discussion on bilanguaging, see Walter D. Mignolo, *Local Histories/Global Designs: Coloniality, Subaltern Knowledges, and Border Thinking* (Princeton, NJ: Princeton University Press, 2012), 250.

19 AnaLouise Keating, "From Borderlands and New Mestizas to Nepantlas and Nepantleras: Anzaldúan Theories for Social Change," *Human Architecture: Journal of the Sociology of Self-Knowledge* 4, no. 3 (2006): 10.

20 Anzaldúa, *Borderlands*, 25. *La herida abierta* is the metaphorical, physical, and economic "open wound" that was formed when border lines were cut and sexualities, genders, and abilities were defined along binary lines, disregarding peoples and how complex we are. We have borders "where the Third World grates against the first and bleeds. And before a scab forms it hemorrhages again, the lifeblood of two worlds merging to form a third country—a border culture" (25).

21 Gloria E. Anzaldúa, "Now Let Us Shift . . . the Path of Conocimiento . . . Inner Work, Public Acts," in *This Bridge We Call Home*, ed. Gloria E. Anzaldúa and AnaLouise Keating (New York: Routledge, 2013), 540. *Conocimiento* as a way of knowing exposes and questions *desconomientos*, blind spots in our knowledge. *Conocimiento* leads to re-membering and healing.

22 Keating, "From Borderlands," 20.

23 *Nepantla* is the Nahuatl word for "in the middle" or the "in-between space" and speaks to the hybrid identity Mexican Americans and other immigrants have. A *nepantlera* is a visionary cultural leader who questions previously accepted worldviews. Living between cultures, the *nepantlera* can have double vision, able to see the perspective of self and other. Claiming the identity of a *nepantlera* is risky and exhausting, as it is an outsider position due to personal and cultural codes constantly clashing. Few travelers are *nepantleras*, but all travelers can learn to be more comfortable in in-between worlds. See Gloria Anzaldúa and AnaLouise Keating, *Light in the Dark / Luz en lo Oscuro: Rewriting Identity, Spirituality, Reality*, Latin America Otherwise: Languages, Empires, Nations: The Gloria Anzaldúa Reader (Durham, NC: Duke University Press, 2015), 2–3; and Max Wolf Valerio, "'Now That You're a White Man': Changing Sex in a Postmodern World—Being, Becoming, and Borders," in *This Bridge We Call Home: Radical Visions for Transformation*, ed. Gloria Anzaldúa and AnaLouise Keating (New York: Routledge, 2002), 242.

24 E. Ann Kaplan, *Looking for the Other: Feminism, Film, and the Imperial Gaze* (New York: Routledge, 1997), 5–6. Kaplan, an English professor interested in how cinema portrays traveling, identifies the gaze as "the male gaze"; there is also the "imperial gaze" (Laura Mulvey), "the disciplinary gaze" (Ella Shohat), and the "orientalist gaze" (Reina Lewis). For the imperial gaze, see Laura Mulvey, "Visual Pleasure and Narrative Cinema," *Screen* 16, no. 3 (1975): 6–18. For the disciplinary gaze, see Ella Shohat, "Imaging Terra Incognita: The Disciplinary Gaze of Empire," *Public Culture* 3, no. 2 (1991): 41–70. For the orientalist gaze, see Reina Lewis, *Gendering Orientalism: Race, Femininity and Representation* (New York: Routledge, 1996), 127.

25 Michael T. Taussig, *Mimesis and Alterity: A Particular History of the Senses* (New York: Routledge, 1993), 20, 35. Taussig, a philosopher, builds his argument on Walter Benjamin's essay "The Work of Art in the Age of Mechanical Reproduction," in *Illuminations: Essays and Reflections* (New York: Schocken Books, 1969),

217–252. In this essay, Benjamin saw a relationship between mimetic desire and a child's relationship with the mother.

26 Kaplan, *Looking for the Other*, 14.

27 The French philosopher and ethicist Emmanuel Levinas describes the gaze as a way of looking that breaks mutuality and operates in a privileged way. See Emmanuel Levinas, *Totality and Infinity: An Essay on Exteriority*, Duquesne Studies. Philosophical Series, vol. 24 (Pittsburgh: Duquesne University Press, 1969), 188. For the gaze within tourism, see John Urry, *The Tourist Gaze: Leisure and Travel in Contemporary Societies* (London: Sage, 1990).

28 Gloria Anzaldúa et al., *The Gloria Anzaldúa Reader* (Durham, NC: Duke University Press, 2009), 79.

29 For I/Thou and I/It relationships and how they inform our relationship with God, see Buber, *I and Thou*.

30 Michael P. Nichols, *The Lost Art of Listening* (New York: Guilford, 1995).

31 Dietrich Bonhoeffer, *Life Together* (New York: Harper & Row, 1954), 83.

32 *Parrhesia* was reintroduced to Western philosophy through Michel Foucault in his 1983 lectures given at the University of California, Berkeley. See Michel Foucault, *Discourse and Truth: The Problematization of Parrhēsia* (Paris: Foucault Archive, 1983); and Michel Foucault and Joseph Pearson, *Fearless Speech* (Los Angeles: Semiotext[e], 2001). It is "speaking boldly," as the word is translated in Acts 14:3.

33 These themes were taken from the Tanzanian artist "Dr. Remmy" Ongala's 1989 album, *Songs of a Poor Man*, released by Real World Studios, a label out of Peter Gabriel's world music organization, WOMAD (World of Music, Arts and Dance).

34 For rewriting history and literary texts, see Ann Rigney and Kiene Brillenburg Wurth, *The Life of Texts: An Introduction to Literary Studies* (Amsterdam: Amsterdam University Press, 2019), 351.

35 Select histories include Bartolome Las Casas and Anthony Pagden, *A Short Account of the Destruction of the Indies* (New York: Penguin, 2004); Charles C. Mann, *1491: New Revelations of the Americas before Columbus*, 1st ed. (New York: Knopf, 2005); Roxanne Dunbar-Ortiz, *An Indigenous Peoples' History of the United States* (Boston: Beacon, 2014); Dee Brown, *Bury My Heart at Wounded Knee: An Indian History of the American West*, 30th anniversary ed. (New York: H. Holt, 2001); John Ehle, *Trail of Tears: The Rise and Fall of the Cherokee Nation*, 1st ed. (New York: Doubleday, 1988); John Hope Franklin and Loren Schweninger, *Runaway Slaves: Rebels on the Plantation* (New York: Oxford University Press, 1999); Daina Ramey Berry and Leslie M. Harris, *Sexuality and Slavery: Reclaiming Intimate Histories in the Americas* (Athens: University of Georgia Press, 2018); Herman L. Bennett, *African Kings and Black Slaves: Sovereignty and Dispossession in the Early Modern Atlantic*, 1st ed. (Philadelphia: University of Pennsylvania Press, 2019); and David W. Blight, *Frederick Douglass: Prophet of Freedom*, 1st Simon & Schuster hardcover ed. (New York: Simon & Schuster, 2018). South African historian Naledi Nomalanga Mkhize wrote a history book for children focused on African perspectives. See Khize Wama Zambezi, *In Africa with Avi and Kumbi* (Johannesburg: Phaphamani Children's Literature, 2018).

36 What follows is a select list of novels and books that were rewritten: The novel *Wide Sargasso Sea* (1966) by Jean Rhys rewrites the classic *Jane Eyre* (1847) by Charlotte Brontë from the perspective of Bertha Mason, the Caribbean wife of Mr. Rochester. He married Jane Eyre, but only after Bertha was diagnosed as insane, kept in the

Rochester home's attic, and died as a ghost. *Wide Sargasso Sea* also writes back from William Shakespeare's *Othello* (1604). Daniel Defoe's *Robinson Crusoe* (1719) was rewritten from a postcolonial perspective by the South African J. M. Coetzee in his *Foe*. See Jean Rhys, *Wide Sargasso Sea* (London: Deutsch, 1966); Charlotte Brontë, *Jane Eyre* (New York: Rinehart, 1950); J. M. Coetzee, *Foe* (New York: Penguin, 1987); and Daniel Defoe, *Robinson Crusoe* (New York: Simon & Brown, 2012). Similarly, Vidiadhar Surajprasad Naipaul's *A Bend in the River* (New York: Clarion, 1979) speaks to Joseph Conrad's *Heart of Darkness* (New York: Palgrave Macmillan, 1996). As another example, Michael Ondaatje's *The English Patient* (London: Bloomsbury, 1992) draws on Rudyard Kipling's *Kim* (Oxford: Oxford University Press, 1987). And Derek Walcott's *Omeros* (New York: Farrar, Straus & Giroux, 1990) rewrites Homer's *The Iliad* (New York: HarperCollins, 2012).

37 For frozen moments, see D. W. Winnicott, "Metapsychological and Clinical Aspects of Regression within the Psycho-analytical Set-Up," in *Through Paediatrics to Psycho-analysis: Collected Papers* (New York: Brunner/Mazel, 1992), 284.

38 For a discussion on touch and porosity, see Luce Irigaray, *An Ethics of Sexual Difference* (Ithaca, NY: Cornell University Press, 1993), 213.

39 Jennie Germann Molz, "Eating Difference: The Cosmopolitan Mobilities of Culinary Tourism," *Space and Culture* 10, no. 1 (2007): 80.

40 Anzaldúa sees the mestiza as a bridge figure, where the bridge is the space between two locals. Travelers can function in similar fashion with their location. "The bridge (boundary between the world you've just left and the one ahead) is both a barrier and point of transformation," Anzaldúa writes. See Anzaldúa, "Now Let Us Shift," 557. She identifies seven "stages," modeled after a religious or ritual pattern, that lead to *conocimiento: un darse cuenta* as a call to adventure; *nepantla* as liminal space; a state of complete darkness that critiques Western ideology; the reconnection of mind, body, and spirit; a new desire for order and meaning; presenting a new script to others; and spiritual activism as a clash between different realities. See also Gloria Anzaldúa and Cherríe Moraga, *This Bridge Called My Back: Writings by Radical Women of Color* (Watertown, MA: Persephone, 1981).

41 Chandra Ford, "Standing on *This Bridge*," in Anzaldúa and Keating, *This Bridge We Call Home*, 312.

42 Shaun Gallagher, *A Neurophenomenology of Awe and Wonder: Towards a Nonreductionist Cognitive Science* (Basingstoke, UK: Palgrave Macmillan, 2015), 6. This book places awe and wonder in the context of space travel by analyzing the in-flight journals of astronauts, adding personal interviews, and engaging in neuroscientific research using electroencephalography (EEG), electrocardiogram (ECG), and functional near-infrared spectroscopy (fNIRS) technologies in simulations of space flight. It brings philosophy and spirituality in conversation with cognitive neuroscience.

43 Kirk Schneider, "The Resurgence of Awe in Psychology: Promise, Hope, and Perils," *Humanistic Psychologist* 45, no. 2 (2017): 103.

44 Dacher Keltner and Jonathan Haidt, "Approaching Awe, a Moral, Spiritual, and Aesthetic Emotion," *Cognition and Emotion* 17, no. 2 (2003): 304–6.

45 Gallagher, *Neurophenomenology*, 6.

46 For coercive confession, see Michel Foucault, *The History of Sexuality*, vol. 1 (New York: Pantheon, 1978). Foucault later acknowledged that not all confessions stem from power relations.

47 For a discussion on reparative intelligence, see Hamman, *Growing Down*, 101.

48 Paulo Freire, *Education for Critical Consciousness* (New York: Continuum, 2005), 131. For Freire, conscientization is reflection that leads to an in-depth understanding of the world, exposing social and political contradictions. Acting is inherent in conscientization.

49 Jaco J. Hamman, "Revisiting Forgiveness as a Pastoral-Theological Problem," *Pastoral Psychology* 61, no. 4 (2012): 435–50.

50 Keltner and Haidt, "Approaching Awe," 303. The authors write, "Vastness refers to anything that is experienced as being much larger than the self, or the self's ordinary level of experience or frame of reference. Vastness is often a matter of simple physical size, but it also involves social size such as fame, authority, or prestige. . . . Accommodation refers to the Piagetian process of adjusting mental structures that cannot assimilate a new experience."

Chapter 4: Caring

1 Jamal, *Justice and Ethics*, xv.

2 Stanford University estimates (December 2019) that the average world citizen is responsible for 5.3 US tons of emissions. The average American is responsible for 18.55 US tons a year. See "Global Carbon Emissions Growth Slows, but Hits Record High," Stanford News, December 3, 2019, https://tinyurl.com/1566ygrb.

3 Proponents of township tours suggest that they create income for local people and bring money directly into the community, they are empowering and an avenue of education for both tourists and hosts, and they shed a much-needed light on issues of urban poverty. Critics of township tourism see the tours as exploitative and voyeuristic, a form of "poverty porn." The divide between these two groups indicates the challenge of addressing sociopolitical and socioeconomic injustices as a traveler. For a discussion of "slum tourism," especially in South Africa and Brazil, see Fabian Frenzel, "Beyond 'Othering': The Political Roots of Slum Tourism," in *Poverty, Power and Ethics in Global Slum Tourism*, ed. Fabian Frenzel, Ko Koens, and Malte Steinbrink (London: Routledge, 2012), 57. For "poverty porn" as an element in slum tours, see Kanika Basu, "Slum Tourism: For the Poor, by the Poor," in Frenzel, Koens, and Steinbrink, *Poverty, Power and Ethics*.

4 John Rawls, *A Theory of Justice* (Cambridge, MA: Belknap, 1971), 303.

5 Päivi Kannisto, "Travelling like Locals: Market Resistance in Long-Term Travel," *Tourism Management* 67 (2018): 302.

6 "Tourism Worldwide 2020," Statista, accessed June 18, 2020, https://www.statista.com/study/9996/tourism-worldwide-statista-dossier/.

7 Jamal, *Justice and Ethics*, 15.

8 For an "environmental bubble" as a "tourist bubble," see Erik Cohen, "Toward a Sociology of International Tourism," *Social Research* 39, no. 1 (1972): 167.

9 The United Nations Sustainable Development Goals for 2030 include six areas of care in the agenda for tourism: having a critical tourism scholarship to foster critical thinking vis-à-vis tourism development; addressing gender and other forms of intersectionality; bringing indigenous and alternative paradigms and values into the tourism conversation; finding alternatives to the progrowth agenda of neoliberal

capitalism; establishing ethical consumption and production in the tourism industry; and, implementing equitable, inclusive, just, ecologically compatible, and thereby sustainable mechanisms of governance. See Karla A. Boluk, Christina T. Cavaliere, and Freya Higgins-Desbiolles, "A Critical Framework for Interrogating the United Nations Sustainable Development Goals 2030 Agenda in Tourism," *Journal of Sustainable Tourism* 27, no. 7 (2019): 847–64.

10 Jamal, *Justice and Ethics*, 169.

11 For a link between care and neoliberalism, see Bruce Rogers-Vaughn, *Caring for Souls in a Neoliberal Age* (New York: Palgrave Macmillan, 2016). See also Adam Kotsko, *Neoliberalism's Demons: On the Political Theology of Late Capital* (Stanford, CA: Stanford University Press, 2018); Ilsup Ahn, *Just Debt: Theology, Ethics, and Neoliberalism* (Waco, TX: Baylor University Press, 2017); and Victoria E. Collins and Dawn L. Rothe, *The Violence of Neoliberalism: Crime, Harm and Inequality* (New York: Taylor & Francis, 2019).

12 Martin Luther King Jr., *A Testament of Hope: The Essential Writings of Martin Luther King, Jr.*, 1st ed. (San Francisco: Harper & Row, 1986), 47.

13 Carolyn Pedwell, "Affective (Self-) Transformations: Empathy, Neoliberalism and International Development," *Feminist Theory* 13, no. 2 (2012): 172. See also Carolyn Pedwell, "Economies of Empathy: Obama, Neoliberalism, and Social Justice," *Environment and Planning D: Society and Space* 30, no. 2 (2012): 280; and Carolyn Pedwell, "De-colonising Empathy: Thinking Affect Transnationally," *Samyukta: A Journal of Women's Studies* 16, no. 1 (2016): 27.

14 Ulrich Beck, *The Cosmopolitan Vision* (Cambridge: Polity, 2006), 6. See also Émilie Crossley, "Cosmopolitan Empathy in Volunteer Tourism: A Psychosocial Perspective," *Tourism Recreation Research* 42, no. 2 (2017): 151.

15 See Jamal, *Justice and Ethics*, 31. Jamal addresses the "trickle-down effect of tourism" as she highlights how ports of call used by cruise ships do not enrich the communities, even as employment opportunities are created by the cruise industry.

16 Pedwell, "Affective (Self-) Transformations," 164.

17 This section draws on the work of University of Minnesota political science professor Joan Tronto. Tronto's work on the ethic of care within gender studies and political theory offers a way through and out of a neoliberal world where care is privatized or denied. See Joan C. Tronto, *Moral Boundaries: A Political Argument for an Ethic of Care* (Hove, UK: Psychology Press, 1993), 102. Carol Gilligan was among the first to call for an ethic of care. See Carol Gilligan, *In a Different Voice: Psychological Theory and Women's Development* (Cambridge, MA: Harvard University Press, 1993), 73. Gilligan argues for an ethic of care and responsibility rather than care tied to rights and rules.

18 Joan C. Tronto, *Caring Democracy: Markets, Equality, and Justice* (New York: NYU Press, 2013), 155. Tronto added "solidarity" and "caring with" as a fifth phase of caring as she built upon *Moral Boundaries* in her book *Caring Democracy*.

19 Tronto, *Moral Boundaries*, 103.

20 Tronto, *Moral Boundaries*, 105.

21 These are the phases Tronto identifies that inform an ethic of care. Joan C. Tronto, "An Ethic of Care," *Generations: Journal of the American Society on Aging* 22, no. 3 (1998): 17. See also Tronto, *Moral Boundaries*, 105–8.

22 For a discussion of the impact of volunteer tourism, see Jim Butcher and Peter Smith, "'Making a Difference': Volunteer Tourism and Development," *Tourism Recreation Research* 35, no. 1 (2010): 27–36.

23 Tronto, *Caring Democracy*, 35.

24 Tronto, *Moral Boundaries*, 122.

25 *Oxford English Dictionary (Online)*, s.v. "Solidarity, N.," accessed February 3, 2021, https://tinyurl.com/1oto5800.

26 Vincent J. Cornelisse et al., "Sexual Safety and HIV Prevention in Travel Medicine: Practical Considerations and New Approaches," *Travel Medicine and Infectious Disease* 28 (2019): 68. See also Milene Gonçalves Lança, João Filipe Marques, and Patrícia Susana Pinto, "Liminality and the Possibilities for Sex and Romance at an International Bike Meeting: A Structural Modeling Approach," *Tourism & Management Studies* 13, no. 1 (2017): 18–26; Kristina Ingemarsdotter Persson et al., "Place and Practice: Sexual Risk Behaviour While Travelling Abroad among Swedish Men Who Have Sex with Men," *Travel Medicine and Infectious Disease* 25 (2018): 58–64; and Summer L. Martins et al., "Pregnancy Prevention on the Fly: An Exploratory Study of Contraceptive Lapse among Young Women Traveling Internationally," *Journal of Women's Health* 28, no. 7 (2019): 951–60.

27 Raoul Bianchi, "Tourism and the Globalisation of Fear: Analysing the Politics of Risk and (in)Security in Global Travel," *Tourism and Hospitality Research* 7, no. 1 (2006): 69.

28 See "The World Factbook," CIA, accessed June 24, 2020, https://www.cia.gov/the-world-factbook/. For the *GINI Index*, which anticipates social disruption the larger the gap between rich and poor, see "World Factbook."

29 The production of care, also referred to as the democratization of care, is a central focus of Joan Tronto's work. She highlights how care is distributed and devalued in neoliberal economies. See Tronto, *Caring Democracy*, 169.

30 Tronto, *Moral Boundaries*, 120. For privileged irresponsibility, also see Tronto, *Caring Democracy*, 41–45.

31 Tronto, *Caring Democracy*, 104.

32 "Politics" as we use it here is found in book 4 of Plato's *The Republic*, accessed January 28, 2021, https://tinyurl.com/y3qqg2fc. For a discussion on Jesus's "politics" using Plato's "shape of the city," see Marcus J. Borg, *Jesus, a New Vision: Spirit, Culture, and the Life of Discipleship*, 1st ed. (San Francisco: Harper & Row, 1987), 129.

33 Joerg Rieger, *Faith on the Road: A Short Theology of Travel & Justice*, Rev. ed. (Downers Grove, IL: InterVarsity, 2015), 15. Whereas Rieger sees "travel" as performed by any person on the move, including displaced individuals, we are exploring traveling as a chosen activity to fulfill one's wanderlust and curiosity. To travel, we said, is to move at the speed of being present to one's experiences, bridging distance and difference through acts of care.

34 Rieger, 16.

35 Rieger, 22.

36 Rieger, 30.

37 Rieger, 39. Rieger quotes theologian Jürgen Moltmann on *christologia viatorum* and *christologia viae*. See Jürgen Moltmann, *The Way of Jesus Christ: Christology in Messianic Dimensions* (Minneapolis: Fortress, 1993), xiv.

38 Rieger, *Faith*, 50.

39 Rieger, 133.

40 Rieger, 109.

41 Jamal, *Justice and Ethics*, 150.
42 Jamal, 157.
43 Jon Martin Denstadli and Knut Veisten, "The Flight Is Valuable Regardless of the Carbon Tax Scheme: A Case Study of Norwegian Leisure Air Travelers," *Tourism Management* 81 (2020): 104–50.
44 Elizabeth Agyeiwaah, Oheneba Akyeampong, and Edem K Amenumey, "International Tourists' Motivations to Choose Homestay: Do Their Socio-demographics Have Any Influence?," *Tourism and Hospitality Research* 13, no. 1 (2013): 21.
45 Quoted in Jamal, *Justice and Ethics*, 157.
46 David A. Fennell and Valerie Sheppard, "Tourism, Animals and the Scales of Justice," *Journal of Sustainable Tourism* 29, nos. 2–3 (2020): 314–35.
47 Jamal, *Justice and Ethics*, 1–2. On being "good," Jamal writes, "To engage in 'good tourism' requires, among other things, being able to identify and address injustices that arise in relation to tourism, directly or indirectly, and for this a holistic view as described further below is important. It applies to every approach and form of tourism that aspires toward being responsible, just, and 'sustainable.'"
48 Pirsig, *Zen*, 223.
49 Pirsig, 164.
50 Pirsig, 267.
51 Pirsig, 289.
52 Tronto, *Moral Boundaries*, 102.
53 Pirsig, *Zen*, 273.
54 Pirsig, 288.
55 Pirsig, 289.
56 Pirsig, 316.
57 Pirsig, 294. We get our word *enthusiasm* from this Greek word.
58 Pirsig, 403.
59 Pirsig, 400.
60 Pirsig, 402.
61 Pirsig, 404.
62 Jacques Derrida, *The Work of Mourning* (Chicago: University of Chicago Press, 2001), 107–8. In *The Work of Mourning*, Derrida remembers his closest friends and colleagues who have died. Many entries are eulogies he wrote.

Chapter 5: Returning

1 Lewis Mitchell et al., "The Geography of Happiness: Connecting Twitter Sentiment and Expression, Demographics, and Objective Characteristics of Place," *PLOS One* 8, no. 5 (2013), accessed January 28, 2021, https://tinyurl.com/y54lf53f.
2 I blogged about this twelve-thousand-mile (nineteen-thousand-kilometer) trip at "Alaska," JacoHamman.com, accessed January 28, 2021, http://www.jacohamman.com/alaska07.htm.
3 In this game, I would lie on my stomach and the girls would tickle me. My task was not to laugh, which would end the game. The game was inspired by reading British psychotherapist Adam Phillips. See Adam Phillips, *On Kissing, Tickling, and Being*

Bored: Psychoanalytic Essays on the Unexamined Life (Cambridge, MA: Harvard University Press, 1993), 9.

4 Depression is a medical illness that needs professional care. The symptoms of depression are sleep disturbances, weight and appetite changes, poor concentration, physical fatigue and slowing down, an especially depressed mood, an inability to enjoy what was life-giving before (anhedonia), feeling of worthlessness and guilt, and suicidal ideation. See "What Is Depression?," American Psychiatric Association, accessed July 19, 2020, https://tinyurl.com/ybbh8w9y.

5 Alves, *Best Chronicles*, 138.

6 See the report "Geospatial Summary of Crash Fatalities," NHTSA, May 2020, https://tinyurl.com/vypb28tn.

7 Pamela Cooper-White, "Shared Wisdom: Use of the Self in Pastoral Care and Counseling (Person, Culture, and Religion Group, American Academy of Religion, November 18, 2005)," *Pastoral Psychology* 55, no. 2 (2006): 235. See also Cooper-White's book, Pamela Cooper-White, *Shared Wisdom: Use of the Self in Pastoral Care and Counseling* (Minneapolis: Fortress, 2004), 48–49.

8 Jody Messler Davies, "Linking the 'Pre-analytic' with the Postclassical: Integration, Dissociation, and the Multiplicity of Unconscious Process," *Contemporary Psychoanalysis* 32, no. 4 (1996): 553–75. The multiple self differentiates, fragments, and reintegrates constantly. Traveling can serve both a fragmenting and a reintegrating function.

9 Davies, 562.

10 Cooper-White builds on Davies's work, saying that the self is not an onion to be peeled layer by layer but really "a child's kaleidoscope in which each glance through the pinhole of a moment in time provides a unique view; a complex organization in which a fixed set of colored, shaped, and textured components rearrange themselves in unique crystalline structures determined by way of infinite pathways of interconnectedness." See Cooper-White, *Shared Wisdom*, 49.

11 Maximiliano E. Korstanje, "Divorcing after Holidays: From Sacredness to Post-vacation Blues Syndrome," in *Global Dynamics in Travel, Tourism, and Hospitality*, ed. Nikolas Pappas and Ilenia Bregoli (Hershey, PA: Business Science Reference, 2016), 71.

12 Sigmund Freud, "Mourning and Melancholia," in *The Standard Edition of the Complete Psychological Works of Sigmund Freud*, ed. James Strachey, vol. 14 (London: Hogarth, 1957), 243.

13 Freud, "Mourning and Melancholia," 251.

14 Sebastian Filep, Jennifer Laing, and Mihaly Csikszentmihalyi, *Positive Tourism* (New York: Taylor & Francis, 2016), xii.

15 Anja Pabel, "Returning Home," in *Tourist Behaviour: The Essential Companion*, ed. Philip L. Pearce (Northampton, MA: Edward Elgar, 2019), 370.

16 Pabel, 379.

17 Tronto, *Moral Boundaries*, 103.

18 Pierre Bourdieu, *Outline of a Theory of Practice* (Cambridge: Cambridge University Press, 1977), 184.

19 Pierre Bourdieu, "The Forms of Capital," in *Handbook of Theory and Research for the Sociology of Education*, ed. J. G. Richardson (Westport, CT: Greenwood, 1986), 242–43. Bourdieu writes,

> Depending on the field in which it functions, and at the cost of the more or less expensive transformations which are the precondition for its efficacy in the field in question, capital can present itself in three fundamental guises: as economic capital, which is immediately and directly convertible into money and may be institutionalized in the form of property rights; as cultural capital, which is convertible, in certain conditions, into economic capital and may be institutionalized in the form of educational qualifications; and as social capital, made up of social obligations ('connections'), which is convertible, in certain conditions, into economic capital and may be institutionalized in the form of a title of nobility.

Bourdieu, 243.

20 Bourdieu, 244.

21 Bourdieu, 245.

22 Bourdieu, 248.

23 Bourdieu, 251.

24 Bourdieu, 252. Bourdieu writes, "The transformation of economic capital into social capital presupposes a specific labor, i.e., an apparently gratuitous expenditure of time, attention, care, concern, which, as is seen in the endeavor to personalize a gift, has the effect of transfiguring the purely monetary import of the exchange and, by the same token, the very meaning of the exchange." For transnational capital as a form of cosmopolitan capital that forms persons to be comfortable with diversity and crossing literal and figurative borders due to study-abroad programs, see Don Weenink, "Pupils' Plans to Study Abroad: Social Reproduction of Transnational Capital?," in *Globalisierung, Bildung und Grenzüberschreitende Mobilität*, ed. Jürgen Gerhards, Hans Silke, and Sören Carlson (Wiesbaden: Springer, 2014), 111.

25 Derrida, *Work of Mourning*, 2.

26 Derrida, 149.

27 Derrida, 3.

28 Derrida, 21.

29 Derrida, 56.

30 Søren Kierkegaard, *Kierkegaard's Writings*, vol. 15, *Upbuilding Discourses in Various Spirits*, trans. Howard V. Hong and Edna H. Hong (Princeton, NJ: Princeton University Press, 2009), 152.

31 Kierkegaard, 167. Kierkegaard continues: "The one human being compares himself with others, the one generation compares itself with the other, and thus the heaped-up pile of comparisons overwhelms a person" (189).

32 Kierkegaard, 182.

33 Melanie Klein, *Envy and Gratitude, and Other Works: 1946–1963*, vol. 3, *The Writings of Melanie Klein* (London: Hogarth Press and the Institute of Psycho-Analysis, 1975). This book was first published in 1957.

34 Klein, 181.

35 Klein, 187.

36 Klein, 180.

37 Klein, 189. The seven deadly sins traditionally are pride, envy, gluttony, greed, lust, sloth, and anger.

38 Klein, 193.

39 Klein, 218–20.

40 Jeroen Nawijn, "Happiness through Vacationing: Just a Temporary Boost or Long-Term Benefits?," *Journal of Happiness Studies* 12, no. 4 (2011): 651.

41 See also Barbara Johnson, *Persons and Things* (Cambridge, MA: Harvard University Press, 2008); John McDargh, *Psychoanalytic Object Relations Theory and the Study of Religion: On Faith and the Imaging of God* (Lanham, MD: University Press of America, 1983); Thomas H. Ogden, *The Matrix of the Mind: Object Relations and the Psychoanalytic Dialogue* (Northvale, NJ: Aronson, 1986); and Ann Belford Ulanov, *Finding Space: Winnicott, God, and Psychic Reality*, 1st ed. (Louisville, KY: Westminster John Knox, 2001). These authors explore our object relationships with persons, things, nature, and the sacred.

42 This reading of the prodigal "younger" son and disgruntled older son draws on Richard Q. Ford, *The Parables of Jesus: Recovering the Art of Listening* (Minneapolis: Fortress, 1997), 90; and Richard Q. Ford, *The Parables of Jesus and the Problems of the World: How Ancient Narratives Comprehend Modern Malaise* (Eugene, OR: Wipf & Stock, 2016), 95.

43 Pirsig, *Zen*, 154.

44 Ford, *Parables of Jesus and the Problems of the World*, 99. Ford shows how "a father and two sons" is a common theme in the Hebrew Scriptures: Abraham with Ishmael and Isaac, Isaac with Esau and Jacob, and Jacob with Joseph and his brothers. In these relationships, the younger son wins out over the older brother(s) and a "robe" plays a significant role. The issue of "the land" is always in the background. In these narratives, too, irony abounds.

45 Alphonso Lingis, *Abuses* (Berkeley: University of California Press, 1994). Lingis's book is a collection of letters he wrote to friends while traveling. This quotation comes from a letter written in Mexico City in which he reflects on French philosopher Marcel Mauss's 1923 work, *The Gift*. Mauss distinguishes a "commodity economy," where objects are disconnected from original owners or producers, and a "gift economy," where gift and giver are inseparable. Mauss would argue that the gifts we receive in traveling are inalienable and thus need to be returned. Gift exchanges lead to interdependence, social cohesion, and solidarity.

46 Trinh T. Minh-ha, "Other Than Myself / My Other Self," in *Travellers' Tales: Narratives of Home and Displacement*, ed. George Robertson et al. (New York: Routledge, 1994), 12. Minh-ha argues persuasively that for many who are forced to travel, the refugees and exiles, writing might be their only home (15).

47 Minh-ha, 21.

48 Minh-ha, 11.

49 Kate Douglas Smith Wiggin, *New Chronicles of Rebecca*, http://www.gutenberg.org/ebooks/1375#download, accessed January 28, 2021. "There *is* a kind of magicness about going far away and then coming back all changed," Wiggin writes. See opening quotation.

50 Postmodern philosophers Gilles Deleuze and Félix Guattari argue that we are "nomads" and thus without a permanent "home," yet we are also rhizomatically attached to each other and all of nature (see chapter 4). See Gilles Deleuze and Félix Guattari, *A Thousand Plateaus: Capitalism and Schizophrenia* (Minneapolis: University of Minnesota Press, 1987), 21, 381.

We Are *Homo Errabundus*: Walking with the Earthroamer

1 The Roman satirist Gaius Lucilius (180–103 BCE) wrote about persons being *Homo errabundus*. The British classical scholar and poet Alfred Edward Housman (1859–1936) reflects on *Homo errabundus* in a 1907 paper on Lucilius. See Alfred E. Housman, *The Classical Papers of A. E. Housman*, ed. F. R. D. Goodyear and J. Diggle, vol. 2, *1897–1914* (Cambridge: Cambridge University Press, 1972), 675.

2 Sallie McFague, *Metaphorical Theology: Models of God in Religious Language* (Philadelphia: Fortress, 1982), 105. See also Sallie McFague, *Models of God: Theology for an Ecological, Nuclear Age* (Philadelphia: Fortress, 1987). In *Models of God*, McFague argues against the monarchical model for God, God as King, and nominates God as Mother or Parent, God as Lover, and God as Friend. Along with McFague, theologians who first pointed us to think in terms of metaphors include Ian G. Barbour, *Myths, Models, and Paradigms: A Comparative Study in Science and Religion* (New York: Harper & Row, 1974); and Janet Martin Soskice, *Metaphor and Religious Language* (New York: Oxford University Press, 1985).

3 Milton H. Erickson, "The Confusion Technique in Hypnosis," *American Journal of Clinical Hypnosis* 6, no. 3 (1964): 183. Erickson, a psychiatrist, used this statement to overcome his patients' cognitive defenses. A person hears the statement and is confused, as the rational mind makes no sense of what has been said. The play on words related to time brings disorientation, which opens the possibility for transformation.

4 Stephen Palmer and Sheila Panchal, *Developmental Coaching: Life Transitions and Generational Perspectives* (New York: Routledge, 2011), 7.

5 Paulo Coelho, *The Pilgrimage: A Contemporary Quest for Ancient Wisdom* (San Francisco: HarperSanFrancisco, 1995), 35. Coelho's novel, *O diário de um Mago*, was first published in 1987.

6 I discovered the word in a discussion of the life of Indian psychoanalyst and philosopher Ashis Nandy. See Ramin Jahanbegloo and Ananya Vajpeyi, *Ashis Nandy: A Life in Dissent*, 1st ed. (New Delhi: Oxford University Press, 2018), 277.

7 For a rich summary of Aristotle on *phronesis* and its relationship to virtues, see Jessica Moss, "'Virtue Makes the Goal Right': Virtue and Phronesis in Aristotle's Ethics," *Phronesis* 56, no. 3 (2011): 204–61. See also Alasdair C. MacIntyre, *After Virtue: A Study in Moral Theory* (Notre Dame, IN: University of Notre Dame Press, 1981), 187–88.

8 The construct comes from the British psychiatrist D. W. Winnicott, whose "ego-orgasm" is similar to Jacques Lacan's "*jouissance*." It may be that Winnicott got his construct from Lacan, who wrote about *jouissance* as early as 1953. See D. W. Winnicott, "The Capacity to Be Alone," in *The Maturational Processes and the Facilitating Environment: Studies in the Theory of Emotional Development* (Madison, CT: International Universities Press, 1994), 34–35. For Lacan's "*jouissance*," see Jacques Lacan and Jacques-Alain Miller, *The Ethics of Psychoanalysis 1959–1960: The Seminar of Jacques Lacan*, book 7, Routledge Classics (London: W. W. Norton, 2008), 191.

9 Masud Kahn, "Ego-Orgasm in Bisexual Love," *International Review of Psycho-Analysis* 1 (1974): 148.

10 Roger Rapoport and Bob Drews, *I Really Should Have Stayed Home: The Worst Journeys from Harare to Eternity* (Oakland, CA: RDR, 2001), xii. This book is the fourth in a series exploring travel horror stories. Fifty-one authors write about their travel horrors.

11 David A. Fennell, "Technology and the Sustainable Tourist in the New Age of Disruption," *Journal of Sustainable Tourism* 29, no. 5 (2020): 1–7.

12 This prayer, amended along the themes we have explored in this book, was prayed by Bishop Woodie White at the 1996 United Methodist Church Conference in Denver. Bishop White, an African American leader and pastor, became the United Methodist Church's bishop for Illinois in 1984 and served not only the United Methodist Church but also the World Council of Churches. Discover more about Bishop White at Council of Bishops: The United Methodist Church, accessed July 11, 2020, https://tinyurl.com/yvcx5sq6. For the prayer in its original form, see United Methodist Church, *Wisconsin Annual Conference: Yearbook and Journal* (Nashville: United Methodist Church, 1996), 280.

BIBLIOGRAPHY

Agyeiwaah, Elizabeth, Oheneba Akyeampong, and Edem K Amenumey. "International Tourists' Motivations to Choose Homestay: Do Their Socio-demographics Have Any Influence?" *Tourism and Hospitality Research* 13, no. 1 (2013): 16–26.

Ahn, Ilsup. *Just Debt: Theology, Ethics, and Neoliberalism*. Waco, TX: Baylor University Press, 2017.

Alarcón, Francisco X. *Laughing Tomatoes and Other Spring Poems = Jitomates Risueños Y Otros Poemas de Primavera*. New York: Children's Book Press, 2012.

Alves, Rubem. *The Best Chronicles of Rubem Alves*. Translated by Glenn A. Cheney. Hanover, CT: New London Librarium, 2017.

Angelou, Maya. *All God's Children Need Traveling Shoes*. 1st Random House ed. New York: Random House, 1986.

Anzaldúa, Gloria. *Borderlands / La Frontera: The New Mestiza*. San Francisco: Aunt Lute, 1987.

———. "Now Let Us Shift . . . the Path of Conocimiento . . . Inner Work, Public Acts." In *This Bridge We Call Home: Radical Visions for Transformation*. Edited by Gloria E. Anzaldúa and AnaLouise Keating, 540–578. New York: Routledge, 2013.

Anzaldúa, Gloria, and AnaLouise Keating. *Light in the Dark / Luz en lo Oscuro: Rewriting Identity, Spirituality, Reality*. Latin America Otherwise: Languages, Empires, Nations: The Gloria Anzaldúa Reader. Durham, NC: Duke University Press, 2015.

———, eds. *This Bridge We Call Home: Radical Visions for Transformation*. New York: Routledge, 2002.

Anzaldúa, Gloria, AnaLouise Keating, Walter D. Mignolo, Irene Silverblatt, and Sonia Saldívar-Hull. *The Gloria Anzaldúa Reader*. Durham, NC: Duke University Press, 2009.

Anzaldúa, Gloria, and Cherríe Moraga. *This Bridge Called My Back: Writings by Radical Women of Color*. Watertown, MA: Persephone, 1981.

Anzaldúa, Gloria E., Simon J. Ortiz, Inéz Hernández-Avila, and Domino Perez. "Speaking across the Divide." *Studies in American Indian Literatures* 15, no. 3/4 (2003): 7–22.

Baker-Fletcher, Karen. "God with Us in the Dust." In *Constructive Theology: A Contemporary Approach to Classical Themes*. Edited by Serene Jones and Paul Lakeland, 188–190. Minneapolis: Fortress, 2005.

Barbour, Ian G. *Myths, Models, and Paradigms: A Comparative Study in Science and Religion*. New York: Harper & Row, 1974.

Basu, Kanika. "Slum Tourism: For the Poor, by the Poor." In *Poverty, Power and Ethics in Global Slum Tourism*. Edited by Fabian Frenzel, Ko Koens, and Malte Steinbrink, 66–82. London: Routledge, 2012.

Bavinck, Herman Johan, John Bolt, and John Vriend. *Reformed Dogmatics*. 4 vols. Grand Rapids, MI: Baker Academic, 2003.

Beck, Ulrich. *The Cosmopolitan Vision*. Cambridge: Polity, 2006.

Benhabib, Seyla. *Situating the Self: Gender, Community, and Postmodernism in Contemporary Ethics*. New York: Routledge, 1992.

Benjamin, Jessica. *The Bonds of Love: Psychoanalysis, Feminism, and the Problem of Domination*. 1st ed. New York: Pantheon, 1988.

———. "Surrender and Self-Discovery: Ghent's Reflections on the 'Force for Growth' and What Impedes It." In *The Collected Papers of Emmanuel Ghent*, 71–76. New York: Routledge, 2018.

———. "Two-Way Streets: Recognition of Difference and the Intersubjective Third." *Differences* 17, no. 1 (2006): 116–146.

Bennett, Herman L. *African Kings and Black Slaves: Sovereignty and Dispossession in the Early Modern Atlantic*. 1st ed. Philadelphia: University of Pennsylvania Press, 2019.

Berkhof, Louis. *Systematic Theology*. Grand Rapids, MI: W. B. Eerdmans, 1996.

Berry, Daina Ramey, and Leslie M. Harris. *Sexuality and Slavery: Reclaiming Intimate Histories in the Americas*. Athens: University of Georgia Press, 2018.

Bhabha, Homi K. *The Location of Culture*. New York: Routledge, 1994.

Bianchi, Raoul. "Tourism and the Globalisation of Fear: Analysing the Politics of Risk and (in)Security in Global Travel." *Tourism and Hospitality Research* 7, no. 1 (2006): 64–74.

Blight, David W. *Frederick Douglass: Prophet of Freedom*. 1st Simon & Schuster hardcover ed. New York: Simon & Schuster, 2018.

Bollas, Christopher. *The Shadow of the Object: Psychoanalysis of the Unthought Known*. London: Free Association, 1987.

Boluk, Karla A., Christina T. Cavaliere, and Freya Higgins-Desbiolles. "A Critical Framework for Interrogating the United Nations Sustainable Development Goals 2030 Agenda in Tourism." *Journal of Sustainable Tourism* 27, no. 7 (2019): 847–864.

Bonhoeffer, Dietrich. *Letters and Papers from Prison*. New York: Macmillan, 1972.

———. *Life Together*. New York: Harper & Row, 1954.

Borg, Marcus J. *Jesus, a New Vision: Spirit, Culture, and the Life of Discipleship*. 1st ed. San Francisco: Harper & Row, 1987.

Bourdieu, Pierre. "The Forms of Capital." In *Handbook of Theory and Research for the Sociology of Education*. Edited by J. G. Richardson, 241–258. Westport, CT: Greenwood, 1986.

———. *Outline of a Theory of Practice*. Cambridge: Cambridge University Press, 1977.

Brontë, Charlotte. *Jane Eyre*. New York: Rinehart, 1950.

Brown, Dee. *Bury My Heart at Wounded Knee: An Indian History of the American West*. 30th anniversary ed. New York: H. Holt, 2001.

Bryant, Rebecca, and Daniel M. Knight. *The Anthropology of the Future*. Cambridge: Cambridge University Press, 2019.

Buber, Martin. *I and Thou*. New York: Charles Scribner's Sons, 1958.

Burnhope, Stephen. *Atonement and the New Perspective: The God of Israel, Covenant, and the Cross*. Eugene, OR: Pickwick, 2018.

Butcher, Jim, and Peter Smith. "'Making a Difference': Volunteer Tourism and Development." *Tourism Recreation Research* 35, no. 1 (2010): 27–36.

Coelho, Paulo. *The Pilgrimage: A Contemporary Quest for Ancient Wisdom*. San Francisco: HarperSanFrancisco, 1995.

Coetzee, J. M. *Foe*. New York: Penguin, 1987.

Cohen, Erik. "Toward a Sociology of International Tourism." *Social Research* 39, no. 1 (1972): 164–182.

Collins, Victoria E., and Dawn L. Rothe. *The Violence of Neoliberalism: Crime, Harm and Inequality*. New York: Taylor & Francis, 2019.

Conrad, Joseph. *Heart of Darkness*. New York: Palgrave Macmillan, 1996.

Cooper-White, Pamela. *Shared Wisdom: Use of the Self in Pastoral Care and Counseling*. Minneapolis: Fortress, 2004.

———. "Shared Wisdom: Use of the Self in Pastoral Care and Counseling (Person, Culture, and Religion Group, American Academy of Religion, November 18, 2005)." *Pastoral Psychology* 55, no. 2 (2006): 233–241.

Cornelisse, Vincent J., Edwina J. Wright, Christopher K. Fairley, and Sarah L. McGuinness. "Sexual Safety and HIV Prevention in Travel Medicine: Practical Considerations and New Approaches." *Travel Medicine and Infectious Disease* 28 (2019): 68–73.

Crossley, Émilie. "Cosmopolitan Empathy in Volunteer Tourism: A Psychosocial Perspective." *Tourism Recreation Research* 42, no. 2 (2017): 150–163.

Davies, Jody Messler. "Linking the 'Pre-analytic' with the Postclassical: Integration, Dissociation, and the Multiplicity of Unconscious Process." *Contemporary Psychoanalysis* 32, no. 4 (1996): 553–576.

de Bernardi, Cecilia. "A Critical Realist Appraisal of Authenticity in Tourism: The Case of the Sámi." *Journal of Critical Realism* 18, no. 4 (2019): 437–452.

de Botton, Alain. *The Art of Travel.* London: Hamish Hamilton, 2002.

de Burgh, Chris. "A Spaceman Came Travelling." *Spanish Train and Other Stories.* London: A&M Records, 1975.

Defoe, Daniel. *Robinson Crusoe.* New York: Simon & Brown, 2012.

Deleuze, Gilles, and Félix Guattari. *A Thousand Plateaus: Capitalism and Schizophrenia.* Minneapolis: University of Minnesota Press, 1987.

Denstadli, Jon Martin, and Knut Veisten. "The Flight Is Valuable Regardless of the Carbon Tax Scheme: A Case Study of Norwegian Leisure Air Travelers." *Tourism Management* 81 (2020): 104150.

DeRoo, Neal. *Futurity in Phenomenology: Promise and Method in Husserl, Levinas, and Derrida.* New York: Fordham University Press, 2013.

Derrida, Jacques. *The Work of Mourning.* Chicago: University of Chicago Press, 2001.

Doi, Takeo. *The Anatomy of Dependence.* Tokyo: Kodansha International, 1973.

———. *The Anatomy of Self: The Individual versus Society.* Tokyo: Kodansha International, 1986.

Dunbar-Ortiz, Roxanne. *An Indigenous Peoples' History of the United States.* Boston: Beacon, 2014.

Ehle, John. *Trail of Tears: The Rise and Fall of the Cherokee Nation.* 1st ed. New York: Doubleday, 1988.

Eigen, Michael. "The Area of Faith in Winnicott, Lacan and Bion." *International Journal of Psycho-Analysis* 62 (1981): 413–433.

Ellis, Albert, and Windy Dryden. *The Practice of Rational Emotive Behavior Therapy.* 2nd ed. New York: Springer, 1997.

Erickson, Milton H. "The Confusion Technique in Hypnosis." *American Journal of Clinical Hypnosis* 6, no. 3 (1964): 183–207.

Erikson, Erik H. *Identity, Youth, and Crisis.* New York: W. W. Norton, 1968.

Fanon, Frantz. *The Wretched of the Earth.* New York: Grove, 1963.

Fennell, David A. "Technology and the Sustainable Tourist in the New Age of Disruption." *Journal of Sustainable Tourism*, 2020, 767–773.

Fennell, David A., and Valerie Sheppard. "Tourism, Animals and the Scales of Justice." *Journal of Sustainable Tourism* 29, nos. 2–3 (2020): 314–335.

Filep, Sebastian, Jennifer Laing, and Mihaly Csikszentmihalyi. *Positive Tourism.* New York: Taylor & Francis, 2016.

Fisk, Anna. *Sex, Sin, and Our Selves: Encounters in Feminist Theology and Contemporary Women's Literature*. Eugene, OR: Wipf & Stock, 2014.

Ford, Richard Q. *The Parables of Jesus and the Problems of the World: How Ancient Narratives Comprehend Modern Malaise*. Eugene, OR: Wipf & Stock, 2016.

———. *The Parables of Jesus: Recovering the Art of Listening*. Minneapolis: Fortress, 1997.

Foucault, Michel. *Discourse and Truth: The Problematization of Parrhēsia*. Paris: Foucault Archive, 1983.

———. *The History of Sexuality*. Vol. 1. New York: Pantheon, 1978.

Foucault, Michel, and Joseph Pearson. *Fearless Speech*. Los Angeles: Semiotext(e), 2001.

Franklin, John Hope, and Loren Schweninger. *Runaway Slaves: Rebels on the Plantation*. New York: Oxford University Press, 1999.

Freire, Paulo. *Education for Critical Consciousness*. New York: Continuum, 2005.

Frenzel, Fabian. "Beyond 'Othering': The Political Roots of Slum Tourism." In *Poverty, Power and Ethics in Global Slum Tourism*. Edited by Fabian Frenzel, Ko Koens, and Malte Steinbrink, 49–65. London: Routledge, 2012.

Freud, Sigmund. "Mourning and Melancholia." In *The Standard Edition of the Complete Psychological Works of Sigmund Freud*. Edited by James Strachey, Vol. 14, 243–258. London: Hogarth, 1957.

Gallagher, Shaun. *A Neurophenomenology of Awe and Wonder: Towards a Non-reductionist Cognitive Science*. Basingstoke, UK: Palgrave Macmillan, 2015.

Germann Molz, Jennie. "Eating Difference: The Cosmopolitan Mobilities of Culinary Tourism." *Space and Culture* 10, no. 1 (2007): 77–93.

Ghent, Emmanuel. "Masochism, Submission, Surrender: Masochism as a Perversion of Surrender." *Contemporary Psychoanalysis* 26, no. 1 (1990): 108–136.

Gilligan, Carol. *In a Different Voice: Psychological Theory and Women's Development*. Cambridge, MA: Harvard University Press, 1993.

Gray, A. M. *Charity in Rabbinic Judaism: Atonement, Rewards, and Righteousness*. New York: Routledge, 2019.

Hamman, Jaco J. *Growing Down: Theology and Human Nature in the Virtual Age*. Waco, TX: Baylor University Press, 2017.

———. *A Play-Full Life: Slowing Down and Seeking Peace*. Cleveland, OH: Pilgrim, 2011.

———. "Revisiting Forgiveness as a Pastoral-Theological Problem." *Pastoral Psychology* 61, no. 4 (2012): 435–450.

Heschel, Abraham Joshua. *God in Search of Man: A Philosophy of Judaism*. New York: Farrar, 1955.

———. *The Prophets*. 1st Perennial Classics ed. New York: Harper Perennial, 2001.

Homer. *The Iliad*. New York: HarperCollins, 2012.

Housman, Alfred E. *The Classical Papers of A. E. Housman*. Vol. 2, *1897–1914*, edited by F. R. D. Goodyear and J. Diggle. Cambridge: Cambridge University Press, 1972.

Hrdy, Sarah Blaffer. *Mother Nature: A History of Mothers, Infants, and Natural Selection*. 1st ed. New York: Pantheon, 1999.

Husserl, Edmund. *Experience and Judgment: Investigations in a Genealogy of Logic*. Translated by Ludwig Landgrebe. London: Routledge, 1973.

Irigaray, Luce. *An Ethics of Sexual Difference*. Ithaca, NY: Cornell University Press, 1993.

Jahanbegloo, Ramin, and Ananya Vajpeyi. *Ashis Nandy: A Life in Dissent*. 1st ed. New Delhi: Oxford University Press, 2018.

Jamal, Tazim. *Justice and Ethics in Tourism*. 1st ed. London: Routledge, 2019.

Johnson, Barbara. *Persons and Things*. Cambridge, MA: Harvard University Press, 2008.

Junior, Nyasha. *Reimagining Hagar: Blackness and Bible*. New York: Oxford University Press, 2019.

Kahn, Masud. "Ego-Orgasm in Bisexual Love." *International Review of Psycho-Analysis* 1 (1974): 143–149.

Kannisto, Päivi. "Travelling like Locals: Market Resistance in Long-Term Travel." *Tourism Management* 67 (2018): 297–306.

Kaplan, E. Ann. *Looking for the Other: Feminism, Film, and the Imperial Gaze*. New York: Routledge, 1997.

Keating, AnaLouise. "From Borderlands and New Mestizas to Nepantlas and Nepantleras: Anzaldúan Theories for Social Change." *Human Architecture: Journal of the Sociology of Self-Knowledge* 4, no. 3 (2006): 5–16.

Keltner, Dacher, and Jonathan Haidt. "Approaching Awe, a Moral, Spiritual, and Aesthetic Emotion." *Cognition and Emotion* 17, no. 2 (2003): 297–314.

Kierkegaard, Søren. *Kierkegaard's Writings*. Vol. 15, *Upbuilding Discourses in Various Spirits*. Translated by Howard V. Hong and Edna H. Hong. Princeton, NJ: Princeton University Press, 2009.

King, Martin Luther, Jr. *A Testament of Hope: The Essential Writings of Martin Luther King, Jr.* 1st ed. San Francisco: Harper & Row, 1986.

Kipling, Rudyard. *Kim*. With an introduction and notes by Alan Sandison. Oxford: Oxford University Press, 1987.

Klein, Melanie. *Envy and Gratitude, and Other Works: 1946–1963*. Vol. 3, *The Writings of Melanie Klein*. London: Hogarth Press and the Institute of Psycho-Analysis, 1975.

Korstanje, Maximiliano E. "Divorcing after Holidays: From Sacredness to Post-vacation Blues Syndrome." In *Global Dynamics in Travel, Tourism, and Hospitality*. Edited by Nikolas Pappas and Ilenia Bregoli, 70–84. Hershey, PA: Business Science Reference, 2016.

Kotsko, Adam. *Neoliberalism's Demons: On the Political Theology of Late Capital*. Stanford: Stanford University Press, 2018.

Kuyper, Abraham. *De Gemeene Gratie in Wetenschap En Kunst*. Vols. 1–3. Leiden, Netherlands: Höveker & Wormser, 1905.

Lacan, Jacques, and Jacques-Alain Miller. *The Ethics of Psychoanalysis, 1959–1960: The Seminar of Jacques Lacan*. Book 7, Routledge Classics. London: W. W. Norton, 2008.

Lacoste, Jean-Yves. "The Phenomenality of Anticipation." In *Phenomenology and Eschatology: Not Yet in the Now*. Edited by Neal DeRoo and John Panteleimon Manoussakis, 15–34. Farnham: Ashgate, 2009.

Lamott, Anne. *Traveling Mercies: Some Thoughts on Faith*. 1st ed. New York: Pantheon, 1999.

Lança, Milene Gonçalves, João Filipe Marques, and Patrícia Susana Pinto. "Liminality and the Possibilities for Sex and Romance at an International Bike Meeting: A Structural Modeling Approach." *Tourism & Management Studies* 13, no. 1 (2017): 18–26.

Laozi. *Dao De Jing: The Book of the Way*. Translated by Moss Roberts. Berkeley: University of California Press, 2001.

Las Casas, Bartolome, and Anthony Pagden. *A Short Account of the Destruction of the Indies*. New York: Penguin, 2004.

Levinas, Emmanuel. *Totality and Infinity: An Essay on Exteriority*. Vol. 24. Pittsburgh: Duquesne University Press, 1969.

Levine, Amy-Jill. *Short Stories by Jesus: The Enigmatic Parables of a Controversial Rabbi*. Nashville: Abingdon, 2014.

Lewis, Reina. *Gendering Orientalism: Race, Femininity and Representation*. New York: Routledge, 1996.

Leys, Ruth. *Newborn Imitation: The Stakes of a Controversy*. Cambridge Elements: Histories of Emotions and the Senses. Edited by Jan Plamper. Cambridge: Cambridge University Press, 2020.

Lingis, Alphonso. *Abuses*. Berkeley: University of California Press, 1994.

MacIntyre, Alasdair C. *After Virtue: A Study in Moral Theory*. Notre Dame, IN: University of Notre Dame Press, 1981.

Mahler, Margaret S., Fred Pine, and Anni Bergman. *The Psychological Birth of the Human Infant: Symbiosis and Individuation*. London: Hutchinson, 1975.

Mann, Charles C. *1491: New Revelations of the Americas before Columbus*. 1st ed. New York: Knopf, 2005.

Martins, Summer L., Wendy L. Hellerstedt, Susan M. Mason, and Sonya S. Brady. "Pregnancy Prevention on the Fly: An Exploratory

Study of Contraceptive Lapse among Young Women Traveling Internationally." *Journal of Women's Health* 28, no. 7 (2019): 951–960.

McDargh, John. *Psychoanalytic Object Relations Theory and the Study of Religion: On Faith and the Imaging of God.* Lanham, MD: University Press of America, 1983.

McFague, Sallie. *Metaphorical Theology: Models of God in Religious Language*. Philadelphia: Fortress, 1982.

———. *Models of God: Theology for an Ecological, Nuclear Age*. Philadelphia: Fortress, 1987.

Meconi, David V. *On Self-Harm, Narcissism, Atonement, and the Vulnerable Christ*. New York: Bloomsbury Academic, 2019.

Mignolo, Walter D. *Local Histories/Global Designs: Coloniality, Subaltern Knowledges, and Border Thinking*. Princeton, NJ: Princeton University Press, 2012.

Minh-ha, Trinh T. "Other Than Myself / My Other Self." In *Travellers' Tales: Narratives of Home and Displacement*. Edited by George Robertson, Melinda Mash, Lisa Tickner, Jon Bird, Barry Curtis, and Tim Putman, 8–26. New York: Routledge, 1994.

Mistral, Gabriela. "The Immigrant Jew/Emigrada Judía." *Mistral Review* 1 (2018): 92–95.

Mitchell, Lewis, Morgan R. Frank, Kameron Decker Harris, Peter Sheridan Dodds, and Christopher M. Danforth. "The Geography of Happiness: Connecting Twitter Sentiment and Expression, Demographics, and Objective Characteristics of Place." *PLOS One* 8, no. 5 (2013). https://tinyurl.com/y54lf53f.

Moltmann, Jürgen. *The Coming of God: Christian Eschatology*. London: SCM, 1996.

———. *The Way of Jesus Christ: Christology in Messianic Dimensions*. Minneapolis: Fortress, 1993.

Moss, Jessica. "'Virtue Makes the Goal Right': Virtue and Phronesis in Aristotle's Ethics." *Phronesis* 56, no. 3 (2011): 204–261.

Mulvey, Laura. "Visual Pleasure and Narrative Cinema." *Screen* 16, no. 3 (1975): 6–18.

Naipaul, Vidiadhar Surajprasad. *A Bend in the River*. New York: Clarion, 1979.

Nandy, Ashis. *The Intimate Enemy: Loss and Recovery of Self under Colonialism*. New Delhi: Oxford University Press, 1988.

Nawijn, Jeroen. "Happiness through Vacationing: Just a Temporary Boost or Long-Term Benefits?" *Journal of Happiness Studies* 12, no. 4 (2011): 651–665.

Nichols, Michael P. *The Lost Art of Listening*. New York: Guilford, 1995.

Ogden, Thomas H. *The Matrix of the Mind: Object Relations and the Psychoanalytic Dialogue*. Northvale, NJ: Aronson, 1986.

Ondaatje, Michael. *The English Patient*. London: Bloomsbury, 1992.

The Oxford English Dictionary (Online). London: Oxford University Press, 2017.

Pabel, Anja. "Returning Home." In *Tourist Behaviour: The Essential Companion*. Edited by Philip L. Pearce, 8–26. Northampton, MA: Edward Elgar, 2019.

Paglia, Camille. *Sexual Personae: Art and Decadence from Nefertiti to Emily Dickinson*. New Haven, CT: Yale University Press, 1990.

Palmer, Stephen, and Sheila Panchal. *Developmental Coaching: Life Transitions and Generational Perspectives*. New York: Routledge, 2011.

Pearce, Philip L. *Tourist Behaviour and the Contemporary World*. Buffalo, NY: Channel View, 2011.

Pedwell, Carolyn. "Affective (Self-) Transformations: Empathy, Neoliberalism and International Development." *Feminist Theory* 13, no. 2 (2012): 163–179.

———. "De-colonising Empathy: Thinking Affect Transnationally." *Samyukta: A Journal of Women's Studies* 16, no. 1 (2016): 27–49.

———. "Economies of Empathy: Obama, Neoliberalism, and Social Justice." *Environment and Planning D: Society and Space* 30, no. 2 (2012): 280–297.

Persson, Kristina Ingemarsdotter, Torsten Berglund, Jakob Bergström, Ronny Tikkanen, Anna Thorson, and Birger Forsberg. "Place and Practice: Sexual Risk Behaviour While Travelling Abroad among Swedish Men Who Have Sex with Men." *Travel Medicine and Infectious Disease* 25 (2018): 58–64.

Peterson, Eugene H. *The Message: The Bible in Contemporary Language*. Colorado Springs: NavPress, 2002.

Phillips, Adam. *Equals*. London: Faber & Faber Ltd., 2002.

———. *Monogamy*. New York: Pantheon, 1996.

———. *On Kissing, Tickling, and Being Bored: Psychoanalytic Essays on the Unexamined Life*. Cambridge, MA: Harvard University Press, 1993.

Phillips, Winfred George. "Rahner's Transcendental Deduction of the Vorgriff." *Thomist: A Speculative Quarterly Review* 56, no. 2 (1992): 257–290.

Pirsig, Robert M. *Zen and the Art of Motorcycle Maintenance: An Inquiry into Values*. Quill ed. New York: Harper Collins, 2009.

Pizer, Barbara. "When the Crunch Is a (K)not: A Crimp in Relational Dialogue." *Psychoanalytic Dialogues* 13, no. 2 (2003): 171–192.

Rahner, Karl. *Foundations of Christian Faith: An Introduction to the Idea of Christianity*. New York: Seabury, 1978.

———. *Spirit in the World*. New York: Continuum, 1994.

Rapoport, Roger, and Bob Drews. *I Really Should Have Stayed Home: The Worst Journeys from Harare to Eternity*. Oakland, CA: RDR, 2001.

Rawls, John. *A Theory of Justice*. Cambridge, MA: Belknap, 1971.

Rhys, Jean. *Wide Sargasso Sea*. London: Deutsch, 1966.

Rieger, Joerg. *Faith on the Road: A Short Theology of Travel & Justice*. Rev. ed. Downers Grove, IL: InterVarsity, 2015.

Rigney, Ann, and Kiene Brillenburg Wurth. *The Life of Texts: An Introduction to Literary Studies*. Amsterdam: Amsterdam University Press, 2019.

Risse, Marielle. "White Knee Socks versus Photojournalist Vests: Distinguishing between Travelers and Tourists." In *Travel Culture: Essays on What Makes Us Go*. Edited by Carol T. Williams, 41–50. Westport, CT: Praeger, 1998.

Rogers-Vaughn, Bruce. *Caring for Souls in a Neoliberal Age*. New York: Palgrave Macmillan, 2016.

Said, Edward W. *Orientalism*. 1st ed. New York: Pantheon, 1978.

———. "Traveling Theory." In *World Literature in Theory*. Edited by David Damrosch, 114–133. New York: Wiley-Blackwell, 2014.

Samstag, Nicholas. "Review of Going Sane: Maps of Happiness by Adam Phillips." *Psychotherapy* 43, no. 3 (2006): 363–365.

Schneider, Kirk. "The Resurgence of Awe in Psychology: Promise, Hope, and Perils." *Humanistic Psychologist* 45, no. 2 (2017): 103–108.

Shohat, Ella. "Imaging Terra Incognita: The Disciplinary Gaze of Empire." *Public Culture* 3, no. 2 (1991): 41–70.

Soskice, Janet Martin. *Metaphor and Religious Language*. New York: Oxford University Press, 1985.

Sperry, Len, and Jonathan J. Sperry. "Psychotherapy and Virtue: Positive Psychology and Spiritual Well-Being." In *Religion, Spirituality, and Positive Psychology: Understanding the Psychological Fruits of Faith*. Edited by Thomas G. Plante, 145–158. Santa Barbara, CA: Praeger, 2012.

Spivak, Gayatri Chakravorty. *Can the Subaltern Speak?* Basingstoke, UK: Macmillan, 1988.

Spurr, David. *The Rhetoric of Empire: Colonial Discourse in Journalism, Travel Writing, and Imperial Administration*. Durham, NC: Duke University Press, 1993.

Stilgoe, John R. *Outside Lies Magic: Regaining History and Awareness in Everyday Places*. New York: Walker, 1998.

Taussig, Michael T. *Mimesis and Alterity: A Particular History of the Senses*. New York: Routledge, 1993.

Tervalon, Melanie, and Jann Murray-Garcia. "Cultural Humility versus Cultural Competence: A Critical Distinction in Defining Physician Training Outcomes in Multicultural Education." *Journal of Health Care for the Poor and Underserved* 9, no. 2 (1998): 117–125.

Trible, Phyllis. *Texts of Terror: Literary-Feminist Readings of Biblical Narratives*. Vol. 13, Overtures to Biblical Theology. Philadelphia: Fortress, 1984.

Tronto, Joan C. *Caring Democracy: Markets, Equality, and Justice*. New York: NYU Press, 2013.

———. "An Ethic of Care." *Generations: Journal of the American Society on Aging* 22, no. 3 (1998): 15–20.

———. *Moral Boundaries: A Political Argument for an Ethic of Care*. Hove, UK: Psychology Press, 1993.

Turner, Victor Witter. *The Ritual Process: Structure and Anti-Structure*. London: Routledge & Kegan Paul, 1969.

Ulanov, Ann Belford. *Finding Space: Winnicott, God, and Psychic Reality*. 1st ed. Louisville, KY: Westminster John Knox, 2001.

United Methodist Church. *Wisconsin Annual Conference: Yearbook and Journal*. Nashville: United Methodist Church, 1996.

Urry, John. *The Tourist Gaze: Leisure and Travel in Contemporary Societies*. London: Sage, 1990.

van Gennep, Arnold. *The Rites of Passage*. London: Routledge & Kegan Paul, 1960.

Van Leeuwen, Mary Stewart. *My Brother's Keeper: What the Social Sciences Do (and Don't) Tell Us about Masculinity*. Downers Grove, IL: InterVarsity, 2002.

Wainaina, Binyavanga. "How to Write about Africa." *Granta* 92, The View from Africa (2005): 92–95.

Walcott, Derek. *Omeros*. New York: Farrar, Straus & Giroux, 1990.

Weenink, Don. "Pupils' Plans to Study Abroad: Social Reproduction of Transnational Capital?" In *Globalisierung, Bildung und Grenzüberschreitende Mobilität*. Edited by Jürgen Gerhards, Hans Silke, and Sören Carlson, 111–126. Wiesbaden: Springer, 2014.

Wiggin, Kate Douglas Smith. *New Chronicles of Rebecca*. (New York: Grossett & Dunlap, 1907); Project Gutenberg, last updated March 10, 2018. http://www.gutenberg.org/ebooks/1375#download.

Winnicott, D. W. "The Capacity to Be Alone." In *The Maturational Processes and the Facilitating Environment: Studies in the Theory of Emotional Development*, 29–36. Madison, CT: International Universities Press, 1994.

———. "Metapsychological and Clinical Aspects of Regression within the Psycho-Analytical Set-Up." In *Through Paediatrics to Psycho-Analysis: Collected Papers*, 278–294. New York: Brunner/Mazel, 1992.

Zambezi, Khize Wama. *In Africa with Avi and Kumbi*. Johannesburg: Phaphamani Children's Literature, 2018.

Zuidema, Sytse U. *Communication and Confrontation: A Philosophical Appraisal and Critique of Modern Society and Contemporary Thought*. Kampen, Netherlands: Kok, 1972.

INDEX

SCRIPTURAL INDEX

The Hebrew Scriptures